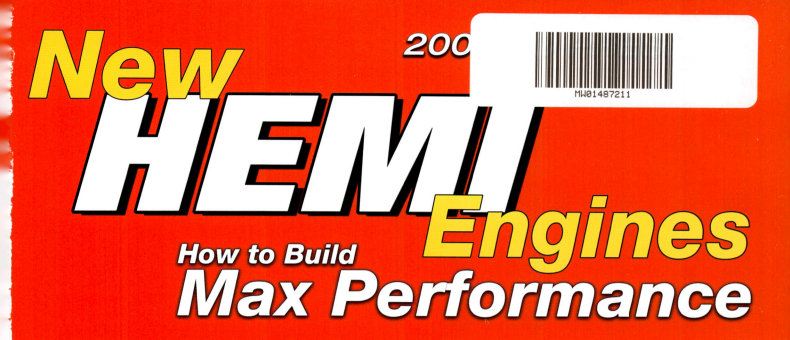

New HEMI Engines

200

How to Build Max Performance

Larry Shepard

S·A DESIGN

CarTech ®

CarTech®

CarTech®, Inc.
838 Lake Street South
Forest Lake, MN 55025
Phone: 651-277-1200 or 800-551-4754
Fax: 651-277-1203
www.cartechbooks.com

© 2017 by Larry Shepard

All rights reserved. No part of this publication may be reproduced or utilized in any form or by any means, electronic or mechanical, including photocopying, recording, or by any information storage and retrieval system, without prior permission from the Publisher. All text, photographs, and artwork are the property of the Author unless otherwise noted or credited.

The information in this work is true and complete to the best of our knowledge. However, all information is presented without any guarantee on the part of the Author or Publisher, who also disclaim any liability incurred in connection with the use of the information and any implied warranties of merchantability or fitness for a particular purpose. Readers are responsible for taking suitable and appropriate safety measures when performing any of the operations or activities described in this work.

All trademarks, trade names, model names and numbers, and other product designations referred to herein are the property of their respective owners and are used solely for identification purposes. This work is a publication of CarTech, Inc., and has not been licensed, approved, sponsored, or endorsed by any other person or entity. The Publisher is not associated with any product, service, or vendor mentioned in this book, and does not endorse the products or services of any vendor mentioned in this book.

Edit by Paul Johnson
Layout by Monica Seiberlich

ISBN 978-1-61325-357-1
Item No. SA404

Library of Congress Cataloging-in-Publication Data
Names: Shepard, Larry, author.
Title: New Hemi engines 2003 to present : how to build max performance / Larry Shepard.
Description: Forest Lake, MN : CarTech, [2017]
Identifiers: LCCN 2017019285 | ISBN 9781613253571
Subjects: LCSH: Chrysler automobile–Motors–Design and construction. | Chrysler automobile–Motors–Modification. | Hemi engine–Design and construction. | Hemi engine–Performance.
Classification: LCC TL215.C55 S484 2017 | DDC 629.25/040288–dc23
LC record available at https://lccn.loc.gov/2017019285

Written, edited, and designed in the U.S.A.
Printed in China
10 9 8 7 6 5 4 3 2

Cover:
A massive Whipple 2.9-liter supercharger feeds this Modern Muscle Extreme 392 Hemi that cranks out 900 hp. It features a full complement of high-performance parts, including 2618 forged pistons and rods, custom performance camshaft grind, custom fabricated intake, custom valvecovers, and other parts. (Photo Courtesy Modern Muscle Extreme)

Title Page:
The 354-ci Gen III Drag Pak engine features a cast-iron block, aluminum heads, and a Whipple supercharger. It powers the Challenger in NHRA Super Stock class against COPO Camaros and Cobra Jet Mustangs.

Back Cover Photos

Top:
In the usual block position, the most notable aspect of the aluminum block is its sleeves. They allow the aluminum block to service many different bore sizes or overbores unlike cast-iron blocks, which are typically limited to .020 to .030 inch over the production bore size.

Middle Left:
Manley makes a rotating assembly for Gen III Hemi engines. The complete package includes a forged crank (4.050 inches is common), H-beam rods, pistons, rings, and bearings. All you need is a block. (Photo Courtesy Manley Performance)

Middle Right:
VVT cams (top) have two grooves in the number-1 journal along with the extra length (about .460 inch) of the number-1 cam journal. Note the small holes drilled into the bottom of the grooves. That is all part of the VVT system.

Bottom:
Most of the engine's valve gear installs on top of the cylinder head. Once the rocker and springs, shafts and hardware are installed, it's pretty crowded.

DISTRIBUTION BY:

Europe
PGUK
63 Hatton Garden
London EC1N 8LE, England
Phone: 020 7061 1980 • Fax: 020 7242 3725
www.pguk.co.uk

Australia
Renniks Publications Ltd.
3/37-39 Green Street
Banksmeadow, NSW 2109, Australia
Phone: 2 9695 7055 • Fax: 2 9695 7355
www.renniks.com

Canada
Login Canada
300 Saulteaux Crescent
Winnipeg, MB, R3J-3T2 Canada
Phone: 800 665 1148 • Fax: 800 665 0103
www.lb.ca

CONTENTS

DEDICATION

To all the Chrysler engineers who designed and built this engine and all
the engine builders and racers who took a chance on a brand-new engine
design in those early years.

ACKNOWLEDGMENTS

The Chrysler/Mopar Generation III Hemi V-8 engine has been in production for almost 14 years, and that's about twice as long as its predecessors. The Hemi Gen I was produced for seven years and the Gen II was offered for eight years. Thousands of the first two Hemi generation engines were built, and now Chrysler has built more than 3.5 million of the Gen III engines. The Gen III went into production after I retired so I had a lot of catching up to do to write this book. Helping with my education was Rob Cunningham of Mancini Racing.

With any new engines, special parts can be an issue, so hats-off to Indy Heads (aluminum intakes), Edelbrock (aluminum heads and intakes), Modern Muscle (aluminum intakes), and Arrow Racing (aluminum blocks) for maintaining and expanding the key performance parts for the Gen III Hemi. Many thanks to all the hard-parts suppliers of cranks, rods, pistons, cams, valvetrain, and gaskets, who have supplied the upgrades to work with the blocks and heads. In addition, I must express my gratitude to all the turners, re-programmers, and computer companies that worked with the new hard parts turning great performance potential into higher and higher horsepower outputs with work-with computers. I want to thank the many manufacturers that display their new hardware at events and shows, such as the Mopar Nats, PRI, and SEMA and their representatives, who are always willing to answer questions and discuss the latest hardware. An extra thanks goes to all manufacturers that supplied photos for this book.

The writing and photography required for an engine book is a lot of work but not anywhere near the amount of work that went into the design and development of the parts originally. Many thanks to Bob Lee and the Chrysler designers for a job well done, plus extra thanks for the early work by Tom Hoover and John Wehrly.

Engine programs have many steps, phases, and milestones and I would like to thank several behind-the-scenes Hemi heroes, including Pat Baer, Bill Hancock, Tim Zuehlke, Jim Szilagyi, and Al Nichols along with Gary Stanton of Stanton Racing Engines. There are too many racers that have helped and are helping with the engine development efforts to name individually but thanks to all.

Extra thanks to Chrysler engineer Joe Kummer for his behind-the-scenes help and research. I would like to thank Bob McSwain of Godfather Racing for his very valuable help and assistance. I would also like to thank Bob Kobylski of Modern Cylinder Head for his patience and helpful assistance.

Many, many thanks to Dave Weber of Modern Muscle and his team for helping me with this book and for showing the way with a new engine program, parts, and development.

I owe the most thanks of all to Dale Matthews of Arrow Racing Engines and his engine builders. I couldn't have done it without them. They all put in lots of time to help me put photos and words together for this book.

Perhaps most of all, I must thank my editor Paul Johnson, for his patience, foresight, and guidance as we took my manuscript and photos and created a readable book.

I would like to give extra special thanks to my wife, Linda, for her steady hand in keeping our household going during this project.

The Gen III Hemi is an all-new engine design introduced in 2003. Although it shares some similarities to previous Hemis, the engine architecture of this engine is substantially different and it carries modern and unique technologies. The Gen III engines evolved from a long line of high-performance and race Hemi engines that also did double duty as production engines.

When designing and building an all-new engine, Chrysler designers had access to leading-edge performance and racing development of Hemi engines being raced for 40-plus years. Add to that baseline the years and years of small-block racing development. This included many variations of cast-iron and aluminum race heads in wedge, canted-valve, and hemi configurations as well as cast-iron and aluminum blocks with tall and short decks. From this mountain of data, designers selected the best features so Chrysler could produce a new generation of engine that was easy to build, durable, lightweight, and responded to performance changes.

It is hard to explain how advanced the Chrysler engineers made this engine when it was introduced in 2003. The Gen III received many awards in the first few years but not much information filtered down to customers. However, the design details of the New Hemi could fill a book on its own. As I researched the specifics of the Gen III engine, I came to realize the amazing job those Chrysler designers did for the performance customer.

In 2011, the production engines took another leap forward in performance. By earlier standards, the 392 was considered close to a race engine with its high compression ratio, high valve lift, big valves, and high-flow ports. This is basically readily available hardware to go racing. For whatever reason, the 392 slid in under the radar.

Gen III Hemi Racing Successes

Just a short "thank you" to those who have gone before. Many, many drivers have raced new Gen III Hemi engines. This little Hall of Fame is for the Hemi engines and racers' accomplishments.

The first production Hemi in 2003 had 340 hp. It is the first production Hemi in over 30 years. A big cheer to Chrysler Engineering.

The 485-hp 6.4 crate engine (from Mopar Performance). The first Hemi crate engine was introduced in

the mid-1990s (Gen II) and this is 20 years later. Welcome to the family.

The 6.2 707-hp Hellcat set the production bar very, very high. An example of the industry's work is Petty's Garage Scat Pack Challenger that makes 720 hp.

BES and Tony Bischoff won the Amsoil Engine Masters Challenge in 2014 with a 400-inch Gen III Hemi that made approximately 695 hp, naturally aspirated.

The 426 naturally aspirated Gen III Hemi used in the NHRA Drag Paks and built by Arrow Racing Engines makes around 780 hp.

Supercharging gets you the big numbers and the Speedkore 2017 Challenger SRT Hellcat makes 1,100 hp.

Another supercharged example of a high-horsepower Hemi is the 1,019-hp Kenne Bell supercharger using 24-psi boost and high-octane race gas.

The Hemi has always put up big numbers, but how about Rob Goss and his Challenger that runs in the

Since the early 2000s there have been many different Gen III Hemi crate engines. The latest 392 Hemi crate makes 485 hp and comes with lots of standard stuff along with many options to help the customer with an install.

X275 drag racing class and holds the 1/4-mile record at 6.85 ET and 204 mph with a 468-inch engine and 27-psi boost?

Specialists

If you build a performance engine, engine specialists will develop it and offer products and services for it. Race competition leads to specialists, and they tend to be local, not at the factory. You could define a "specialist" as an expert on engines and he builds lots of them, and typically he has been doing it for years and is experienced. There are engine specialists for the 354/392 Gen I Hemi, for the 426 Gen II Hemi, for the 440 wedge, and for the 340/360 small-blocks. However, there are not many Gen III experts, which is one of the main reasons for writing this book.

The Gen III started slowly in 2003. The Chrysler bankruptcy in 2008 and the downsizing years leading up to it were a key reason. It started to gain some momentum in 2009–2010 with the Challenger Drag Pak cars and with the SRT car models. The big jump in hard parts came in 2011 with the 392/6.4, and the specialists noticed.

Crate engine builders were a big help after Chrysler came out of bankruptcy and did a great job making performance engines available to customers. They were the first companies to sell engines to performance customers, and these engines were used for racers, engine swapping, custom cars, and other applications. If you want to build a race car, you need a race engine, spare engine, test engine, and development engine and crate engine assemblies is a good source.

One of the performance industry's advantages (in relation to the

The 426 Drag Pak is somewhat of a crate engine and is one of the more impressive versions with its extra-tall intake manifold. Arrow Racing Engines offers other versions.

Gen III engine family) is that the basic parts interchange from one version to another easily thanks to retaining one basic block and head design, so far. Consistency also helps aftermarket parts manufacturers. Another advantage is that the Gen III responds easily to new high-performance hardware.

Hot Rodding

In late 1950s and early 1960s hot rodding began with engine swapping: install a V-8 in place of a 4 or 6; install a big V-8 in place of a small V-8. The Magnum engines, early 440, or 426 Gen II was often hopped up. Gen III seems to be back to the basics of engine swapping: you can install it in anything.

Over the years, hot rodding has changed. Tweaking hardware, such as cams, intakes carbs and heads, has been traded for tweaking electronics, primarily the ECM. Basically, this requires MPI engines at every step. Although engine swapping is not the

main focus of this book, Schumacher makes mounts that allow the Gen III to be swapped easily into many vehicles.

Advanced Technology

Although I discuss high-tech subjects such as MDS, VVT, and active intake manifold, they are not my main focus. To the performance customer, MPI seems to be the technology standard and all Gen III engines come with MPI.

The basic high-tech MPI system was introduced on the 2.2-liter 4-cylinder turbo in 1984 and 100 percent of production on 1992–2003 5.2 and 5.9 Magnum V-8s, so it is not new. In many cases, performance customers who use the engine in a swapping project prefer carburetors. I think it is related to cost.

Another new trend that also may be related to cost is the greatly increased popularity of throttle-body-injection fuel injection systems. These systems are generally more advanced than the early TBI systems, but the electronics are much, much better.

In the mid-to-late 1990s and early 2000s, OEM factories helped with ECM reprogramming (called reflashing) and offered ECMs and basic assistance. Breaking the ECM codes was difficult. Then came the code breakers, and they offered this service and now the aftermarket has work-with, handheld units that can do almost everything.

Engine Development

In researching and writing this book, one thing became obvious: The engine development of the Gen III was just starting. All the development done for the 426 Gen II, 440, and 340 that helped performance customers put together high-performance packages had not been done for the Gen III. Hardware, such as the cam and valvetrain, are moving toward higher valve lifts with a bigger valvespring. Prototypes for high-lift cams, rockers, and mechanical cams are in development but not in the catalog yet. Hydraulic cams with .625- and .650-inch lift

were unheard of in the muscle car era when .600+ lift cams were common but they were all mechanical.

Don't sell production engineers short; they built the 485-hp 392 and 707-hp Hellcat, which are great building blocks for any performance project. These high-performance production parts have not been easy to obtain, so the aftermarket is starting to make them. Today, the aftermarket is leading the way (in parts), but production engineers keep raising the bar, moving the production engine up in output, new parts, and new hardware. When it comes to horsepower ratings, the 5.7 had a modest beginning at 345 hp, which is now up to 385 hp. The 485 hp rating on the 392/6.4 is just plain impressive for an engine that is fully emission certified and warranted.

Basic Topics

When I started to write this book there was not much performance info available. This book is not intended to be a service book, but I tried to cover some of the tricks because there is no other source. I had to put a stake in the ground and write knowing that

The latest performance package is the special circle-track package based on the aluminum block designed for the Canadian circle-track racing series, which uses a distributor (left) and a 390-cfm carburetor (not shown).

some questions would be answered after the book went to press because things are changing fast. This is good for the performance customer.

I began trying to cover everything, but parts are being developed and introduced quickly now. They come out too soon to obtain tests and dyno comparisons for all the new packages and hardware, but I have included most of the 2015–2016 hardware.

Chapter Overview

Over the course of 13 chapters, I will try to cover everything in the engine but nothing more. The hardware discussions are covered as follows:

Cylinder blocks: short, stiff, skirted, and cross-bolted; easy to swap, actually a small-block in size, not in horsepower; best engine for swapping.

Heads: high-flow, emissions, economy; big installed height on the 6.4/392 is a big advantage; it may be the best production head ever.

Cams: the big 392/6.4 at .571-inch lift; cam lift as high as early 426 Gen II race engines (.590 inch) from mid- to late 1960s in high-volume production and full emissions and warranty; the 392 hardware is full race on the street.

Intakes: MPI, smart (dual runner) long runner for torque; short runner for high performance.

Cranks, rods, and pistons: lightweight, low inertia; the perfect building block.

Compression ratio: In the mid-1990s, MPI V-8 engines (5.9) had a compression ratio (CR) of 9.5:1; the 392/6.4 had 10.7:1.

Computer electronics: Known by many names, including ECM, PCM re-flash, or re-program.

A BRIEF HEMI HISTORY

Chrysler Corporation has been closely tied to the Hemi engine design for many years. The Hemi story begins with the Xi 2220 military aircraft project in 1944–1945, which was a V-16 engine designed and developed by Chrysler engineers and produced more than 2,500 hp. Chrysler incorporated this engine technology into the first production Hemi for passenger cars, and in 1951 it became the first member of the Hemi family.

The first production Hemi 331 was also the first Gen I engine, but this moniker was not used at this time. "Hemi" is a shortened version of its general description, which is hemispherical, but that's too long. So, all of these engines are just called hemis. The term "hemispherical" refers to the shape of the combustion chamber in the cylinder head. That shape has always been a key feature in defining basic engine designs, such as wedge or flathead. Although there are many wedge chamber shapes, the Hemi is based on a sphere. The other key aspect of these hemi chambers is that the valves are opposed. This means that the intake valve is on one side of the chamber and the exhaust valve is on the other side, rather than being next to each other as they are in a wedge head.

After the 331 was introduced, Chrysler released larger versions; the 354 was next and then the famous 392, which was the first true Top Fuel engine. It was the first engine that was able to control the power created by racing fuel, not gasoline but alcohol and nitro-methane, and superchargers. It became the engine to beat in NHRA's top class. Production of the Chrysler Hemis (the 392) ended in 1958.

In 1964, the soon-to-be famous 426 Hemi was released, and it has been racing ever since. Rated at a conservative 425 hp, the 426 was based on Chrysler's big-block called the B-engine or, more specifically, the RB-engine. Tom Hoover and his engineering team designed and developed it and it remained in production through 1971. During eight years of production, Chrysler built only 10,669 426 Hemi engines.

The Gen III Hemi was introduced in 2003. To date there have been four production displacements (5.7, 6.1, 6.4, and 6.2) plus

The 354 Drag Pak engine uses a Whipple supercharged on top; it is black with ribs. The billet fuel rail allows increased fuel flow. The supercharger drive is the serpentine belt on the right. Note the rear sump on the aluminum oil pan.

two non-production displacements (426 and 354). The 354 is a super-charged Hemi used only in the Challenger Drag Pak, and the 426 is an aluminum-block version that is used in crate engines sold by Mopar Performance and Arrow Racing Engines. It's also used in the Challenger Drag Pak naturally aspirated models.

Production Engines

By the end of the 2016 model year, Chrysler had manufactured more than 3.5 million Gen III Hemis. Therefore, the New Hemi production volume, performance per cube, and model variations outmatch the Gen I and II Hemis that came before it. Compare this huge number to the just over 10,000 426 Gen II Hemis that were built in eight years of production.

The original 5.7 or 345-ci Hemi was introduced at 345 hp, which has increased to around 366 to 390 hp with the Eagle package, which was introduced in 2009. The newer and larger 6.1 version was introduced at 425 hp, which matches the original Street Hemi Gen II power rating in 1966–1971 models with 372 versus 426 ci. Keep in mind that the SAE engine/horsepower rating system was much stricter in 2010 than it was in the late 1960s.

In 2011 the larger 6.4 or 392 Hemi was introduced, rated at 470 hp. The original Gen I 392 in 1958 made right around 1 hp per cubic inch (hp/ci). All of this was topped in 2014–2015 with the introduction of the 6.2 Hellcat supercharged engine, which was rated at 707 hp. It is Chrysler's first supercharged production engine and has the highest horsepower rating of *any* production engine to date.

5.7-Liter

The 5.7 engine, or 345 ci, which is just slightly larger than the 1968–1973 340 small-block engine uses a 3.917-inch bore, sometimes rounded up to 3.92 inches. The stroke is 3.58 inches, similar to the 360 and 5.9 small-blocks. The original power rating of 345 hp in 2003 yielded 1 hp/ci, which is excellent for any V-8 production engine. The original Gen II Hemi was 1 hp under at 425 hp from 426 ci.

The original 5.7 engine was produced from 2003 through 2008 and then the 5.7 Eagle was introduced in 2009. The high-performance Eagle package is based on a new, high-flow cylinder head with larger ports and bigger valves. The Eagle intake valves are 2.05 inches compared to the standard 2.00 inches. The valves are also about .300 inch longer and the installed height is increased to 1.99 inch, up from 1.81 inches. The bigger ports and bigger intake valves allows the Eagle intake ports to flow about 40 cfm more than standard and the Eagle version makes about 30 hp more (366 to 390) that the original.

6.1-Liter

Introduced in 2006, the 6.1, or 372-ci, Hemi engine was produced

through 2011. A big-bore version of the 5.7, it has a 4.055-inch bore that's sometimes rounded up to 4.06 inches. This bigger engine shares the 3.58-inch stroke with the 5.7. It was originally rated at 425 hp or 1.14 hp/ci. The 6.1 has slightly larger valves (2.075 versus 2.05 inches) compared to the 5.7 Eagle. The valves are slightly longer and the installed spring height is slightly higher (1.87-inch versus 1.81 inches).

A much bigger camshaft is the key to this engine's performance package. The 6.1's valve lift increased almost .100 inch over the 5.7 standard and Eagle cams, and that's actually .472 versus .571 inch; on the exhaust side is .460 versus .551 inch. The advertised duration picked up about 20 degrees (260 versus 283). At .571 valve lift, this has to be one of the largest (highest lift) cams ever used in a production engine.

The previous high-lift high-performance production engines, such as the 426 Hemi, 340, 440-6, etc., used cams with lifts around .450 to .475 inch. High-performance aftermarket cams for these engines tended to peak at around .510 inch, so by comparison, the .571-inch lift is impressive.

Most Gen III Hemis look alike after they are assembled, with the exception of supercharged Hemis. This 6.1-liter version has the typical round (beer-barrel) intake manifold and single, forward-facing, large 80-mm throttle body. The coil mounts directly on top of the plug and sits on top of the valvecovers. It has no distributor and the exhaust looks like a shorty header. The fuel injectors mount in the intake manifold next to the head's intake port, one per runner. The fuel rail mounts on top and attaches to the intake.

6.4-Liter

The 6.4 version of the Gen III Hemi has somewhat of an identity crisis. It was originally called the 392, then the 6.4, and then back to the 392. It has the same displacement either way. This engine package, called the Apache, is rated at 470 hp or 490 hp, which is 1.20 hp/ci. It has the largest valves to date from the production Hemi engines with 2.138-inch intakes and 1.654-inch exhausts. In addition, it features the raised valvetrain with the tallest installed spring heights at 2.051-inch intake and 2.016-inch exhaust.

The big-port Apache head flows almost 340 cfm. The beer-barrel-shaped intake manifold is plastic (lightweight) and features an angled, single inlet. It also has variable valve timing (VVT).

One interesting feature of the Gen III Hemi is that they all have windage trays but they are part of the oil pan gasket rather than a separate tray with an oil pan gasket on top and bottom.

6.2 Hellcat

The supercharged Hellcat engine is similar to the 6.1 and 6.4 Hemis. It uses a cast-iron block with a 4.09-inch bore, the same as the 6.4/392. It also uses

This is the Hellcat supercharged Hemi out of the engine compartment. The large aluminum box at the top-center is the IHI Supercharger. It also has the angled inlet, which is not visible at the upper right. The "supercharged Hemi" cover sits on top of the actual orange valvecovers. When this engine is installed in a car, the supercharger's cover makes it difficult to see the valvecovers.

The 5.7 Hemi (shown) is similar to the 6.1 with the large round intake and the straight-ahead single throttle body. The 5.7 Eagle and the 392 are plastic and the 6.1 intake is aluminum. The black valvecovers on all three versions are plastic. (Photo Courtesy Modern Muscle Extreme)

the 6.1's 3.58-inch stroke. This makes the engine a 6.2 version, or 378 ci. The engine is rated at 707 hp and 650 ft-lbs of torque. The Hellcat uses a similar cam to the 392; .571-inch lift, but it has 8 degrees less intake duration and 16 more degrees of exhaust duration.

IHI Corporation builds the twin-screw supercharger that displaces 2,380 cc. The boost pressure

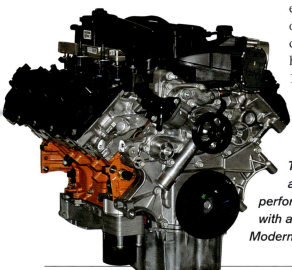

The 392 (6.4) Hemi pumps out 485 hp and 475 ft-lbs of torque for class-leading performance. It uses a single throttle body with a black plastic intake. (Photo Courtesy Modern Muscle Extreme)

is electronically regulated to 11.6 psi. Hellcat blocks are painted orange. A serpentine drive belt drives the supercharger. The heads are aluminum but use smaller valves than the 392 engine. It maintains the long valves and tall installed height.

426 Crate Engine

The 426 Gen III engine uses an aluminum block with 4.125-inch bores and a 4.00-inch crank to gain the 426 ci. As a crate engine, it is rated at 540 hp, or 1.26 hp/ci. This means that the engine isn't just larger, it is also more efficient. The rotating assembly features a forged crank, forged H-beam connecting rods, and forged pistons.

The Drag Pak version increases the compression ratio to 14.9:1. The aluminum head has bigger 2.20-inch

The 426 crate engine is the biggest and baddest of the factory crate engines to date. Shown is the current Drag Pak naturally aspirated engine. It features an aluminum block, aluminum heads, an aluminum very-high-rise tall single-plane intake manifold with the Wilson 4 x 2-inch throttle body, and billet fuel rails.

intake and 1.68-inch exhaust valves. The Drag Pak version also uses 1.80 to 1.85 high-ratio rocker arms made by T&D; the standard ratio is 1.6. The cam is a mechanical roller with .675-inch total valve lift and a hydraulic roller in the standard crate engine.

These crate engines are readily available from Arrow Racing Engines and others.

354 Drag Pak

The 354 Drag Pak Hemi is a unique engine because it is designed specifically for NHRA Super Stock drag racing and not actually a production engine. The sanctioning body mandates the displacement and the supercharger is based on the class (other supercharged engines), but it is similar to the Hellcat, just slightly smaller and uses a different supercharger.

The 354 supercharger is a 2.9 Whipple twin-screw design with an aluminum intake manifold. It also uses a 109-mm billet throttle body. The block is cast iron based on the 6.1 base engine and uses the 4.055- or 4.060-inch bore of the original 6.1 engine. It uses a 3.40-inch forged crank to gain the proper 354 ci. The 354 Hemi uses a mechanical roller cam with 1.6:1-ratio T&D steel rocker arms with .675-inch valve lift. The crank is forged and the rods are forged H-beam designs. The aluminum head uses 2.145-inch intake and 1.660-inch exhaust valves.

Custom Engine Packages

After a few years of production several Mopar engine specialists, including Arrington, put together special engine packages typically based on the 392 production engine, which has been available since 2011. In general, they use a supercharger similar to the Whipple or Magnuson or similar belt-driven unit. These special engine packages are usually installed in production cars such as the Challenger, Charger, and Chrysler 300.

There are several aftermarket engine packages for the Gen III Hemi production engines. Arrington developed this 392; it is rated at 720 hp with 740 ft-lbs of torque and is based on the Magna-Charger. It is also black and has the angled inlet.

Car and Truck Models

Collectors like limited-production cars and with 3.5 million produced, the Gen III Hemis are in a class by themselves; desired but not limited. Although there are millions of Gen III production vehicles, some limited-production models are worth noting. Collectors may not be interested in the millions of Gen III production vehicles, but the "hot rod" and racing industries are interested. These cars give them a readily available Hemi engine that makes a lot of horsepower and easily responds to modifications with even more horsepower.

5.7 Trucks

The first two years of Gen III Hemi production consisted mainly of trucks, lots of trucks. In 2005, cars and Jeeps were added to the production model list. The trucks have continued using the 5.7 Hemi. Therefore, the 5.7 Eagle, an upgraded engine package introduced in 2009, was a big performance upgrade. So far, the bigger Hemi engines (6.1, 6.4, and Hellcat) have not found their way into production trucks. The new (since 2014) 6.4 trucks, called the Big Gas engine, could be a collector vehicle.

6.1 Cars

The 6.1 Hemi engine was introduced in 2006 and was available in models such as the Charger, Magnum, Chrysler 300, and Jeep Grand Cherokee. Many of these vehicles were offered in a performance package called the SRT8. This engine option generally came with a special hood scoop, but there are many variations. Typically, the specials have the SRT8 logo on the right rear panel near the taillight.

SRT8 Grand Cherokee

In 2009 the SRT8 Jeep Grand Cherokee equipped with a 6.1 Hemi engine hit the market, and in 2011, the 392/6.4 was installed. The vehicle has all-wheel drive and weighs about 5,300 pounds; and that's about 1,000 pounds heavier than the Challenger performance cars. In the last few years, these SRT8 Jeeps (2012–2016) run 0-60 in about 4.5 seconds and the quarter-mile in about 13.2 seconds at about 105 mph according to Car and Driver. In the late 1960s and early 1970s, a stock production Hemi car owner would be very happy with these numbers.

The oil filter on the Jeep Hemi engines is relocated at about 90-degrees rearward because of the all-wheel-drive hardware typical of the Jeep vehicle.

Optional Performance Packages for Car Models

Because Gen III Hemi engines were introduced about 14 years ago, many special performance packages have been produced with unique model designations including SRT, Drag Pak, Scat Pack, Hellcat, and

Many of the limited-production special packages based on the Gen III Hemi engine feature a forward-facing hood scoop to provide cold air to the engine. The popular 392 Scat Pack has a 1970 Trans Am Challenger style of scoop and hood.

Demon. The details of these special models may change from year to year. I discuss some of the highlights here.

392 Scat Pack

Available for the past few years, the Scat Pack, a special option on Challenger models, is typically offered for the 392 engine. A hood scoop is part of the package, but several styles are available. The base car is usually a stock Challenger. Typically, these cars weigh the same as the other Challengers at about 4,400 pounds. In 2015, Car and Driver magazine tested the 392 Scat Pack and found it went 0-60 mph in 4.1 seconds and the quarter-mile at 12.4 seconds and 115 mph. That's pretty fast for a stock production car.

Drag Pak

The Drag Pak program started about eight years ago and it changes each year. The base vehicle is a production Challenger with lots of weight removed and certain safety equipment added along with special drivetrain hardware. They are designed to be drag race cars, more specifically Super Stock cars.

They are typically built with one or two engine options. Although the 5.7, 6.1, and 392 have been used in past Drag Paks as engine options, currently two engines are available, a

426 aluminum-block, naturally aspirated engine with a very large Wilson, 4 x 2.0-inch throttle body and a 354 Whipple supercharged with a 109-mm throttle body.

707-hp Hellcat

The top dog in the production horsepower competition at 707 hp, the Hellcat is Chrysler's first supercharged production engine. The Hellcat engine came in the Challenger and Charger models in 2015 and 2016. These cars typically weigh about 4,400 pounds.

Motor Trend magazine tested a Challenger Hellcat that ran 0-60 in 3.7 seconds and the quarter-mile in 11.7 seconds at 125.4 mph. With drag slicks, these cars have gone 10.80 seconds in the quarter. Independent testing has shown that the Challenger Hellcat can travel at 199 mph at the top end. The Charger is slightly more aerodynamic and should go 205 mph with the same engine. The Charger is also about 200 pounds lighter than the Challenger but no tests on that specific model are available yet. The 840-hp Demon is new!

Custom Production Cars

Several aftermarket specialty companies, such as Petty Enterprises,

The 707-hp IHI twin-screw supercharger looks unique in the engine compartment. The supercharger is aluminum and ribbed. The serpentine front drive belt (center at the bottom, next to the throttle body) drives the supercharger. The "Hemi" logo cover sits on top of the standard valvecover; it's orange on Hellcats.

The custom flat and smooth firewall indicates a custom or show car. The pink valve covers match the body color. The single-plane intake manifold indicates fuel injection and a 4-barrel throttle body.

The Gen III Hemi is actually a small engine similar to the Mopar A-engine (340/360) small-blocks and much smaller than the Mopar big-blocks including the Gen II. It fits nicely into the small engine compartment offered by the mid-1930s street rod. The 5.7 Gen III Hemi is trimmed in "body-color" (orange) but it does not use the stock "beer-barrel" intake. Based on the 1970-style air cleaner, it could be a 2 x 4 barrel throttle body set-up (intake by Edelbrock) or a 3 x 2 barrel throttle body set-up (intake by Indy Heads)

have put together special cars typically using a special engine package. Although these special cars may have unique body ornamentation, such as spoilers and hood scoops along with special wheels and tires (bigger), the main focus is a very powerful engine package typically based on the 392 engine.

Swapping

The use of the Gen III Hemi in engine-swapping applications has been somewhat slow to get started because taking a brand-new engine out of a brand-new car can be an expensive way to build a street rod or street machine. These engines are now more than 13 years old and having 3.5 million produced along with the crate engines means that there are lots of ways to get the basics.

The most popular swaps are putting the Gen III into a typical muscle car (1962 through 1976), but the older street rods are becoming more common, along with custom and show cars. The big, wide valvecovers have always been a plus for engine compartment appearance and the basic

high technology of the stock Gen III engine gives any project an advantage.

Recently, Schumacher Creative Services offered bolt-in motor mounts for installing Gen III Hemis into the typical A-, B-, and E-Body muscle car from the late 1960s and early 1970s. Milodon even offers special oil pans for this engine-swap application.

Crate Engine

Crate engine options are constantly evolving. The 5.7, 6.1, and 6.4/392 have all been offered as crate engines dating back to the initial introduction in 2004. Specific details of a crate engine changes are frequently based on customer demand and requests. Today there are 5.7, 392, and 426 crate engines available.

The 426 is rated at 540 hp, based on the 6.1 aluminum intake manifold and large throttle body. There is also a higher output version used in Drag Pak cars, which uses a larger 4-barrel throttle body (Wilson) and a very high-rise single-plane intake manifold (Mopar Performance) in place of the "beer barrel" round intake.

Chrysler dealers and many engine builders offer their own crate engines, which may vary from some of the factory offerings in specific hardware and performance ratings. Currently, Mopar Performance does not offer a Hellcat crate engine. However, several performance dealers offer supercharged 392 engines using the Magnuson supercharger. They look very similar to the Hellcat.

This is the newest naturally aspirated 426 crate engine for Drag Pak cars. The throttle body on top is a Wilson with four 2-inch throttles. The high-flow, billet fuel rails and high-flow injectors are all part of the basic package. Note that the high-rise single-plane intake is very tall but is a one-piece casting. The earlier model was somewhat shorter and used a thick spacer (about 2 inches) between the throttle body and the manifold.

BLOCKS

The third-generation Hemi differed from its immediate predecessor, the small-block Magnum 5.2/5.9 (318/360s). The camshaft was repositioned higher in the block, and had two particular benefits: The valvetrain interior was decreased and a simpler rocker arm arrangement was used. In addition, Chrysler designed hemi-shaped combustion chambers, used twin spark plugs, and a low volume, and a small-surface high-efficiency squish area for greater overall performance. All of these features were incorporated on the first Gen III Hemi, the 5.7, and the heads flowed far better than the competing GM LS.

The stout block easily supported 340 hp, and in the following years it eventually supported the 6.2 supercharged Hellcat engine and its 707 hp. The Gen III was indeed a new engine because it incorporated VVT, and thus had a different oiling system and other related equipment. The Gen III Hemi 5.7 was introduced in 2003 and Chrysler has built more than 3.5 million in multiple versions; production is still going strong.

Basic Design

The Gen III Hemi engine design architecture fits in the general category of 90-degree V-8 similar to its predecessors. Being a Hemi engine also means that the basic valve arrangement has the intake valve on one side of the combustion chamber and the exhaust valve on the opposite side rather than next to each other in the popular wedge-head designs used on many V-8 engines, including the early Mopar small-blocks and

Although most Gen III Hemi blocks tend to look similar, the aluminum block used in the 426 Drag Pack engines is more attractive unpainted than the cast-iron versions, which are often painted black. All production Gen III Hemi engines use multi-point injection (MPI) and a round, beer-barrel-shaped intake manifold, this max-performance version uses an extra-high-rise intake manifold with a 4-barrel throttle body and larger, high-performance fuel rails.

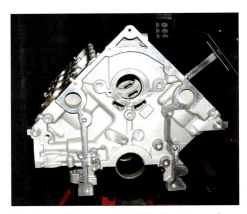

Perhaps the most visible feature of Gen III blocks is the high "point" of the front of the block, nicknamed the birdhouse. Not as obvious is the skirted block, which extends below the main cap centerline similar to that of Gen II blocks.

This main caps view shows the skirted block and the cross-bolted mains: two vertical bolts and two cross-bolted bolts.

Hemi Engine Block Specifications					
	5.7	**6.1**	**6.4/392**	**6.2**	**426**
Years	2003–2016	2006–2010	2011–2016	2014–2016	2013–2016
Weight (pounds, approximate)	195	195	195	195	100
Cylinder Bore (inches)	3.917	4.055	4.09	4.09	4.125
Cylinder Bore Spacing (inches)	4.46	4.46	4.46	4.46	4.46
Main Bearing Bore Diameter	*	*	*	*	*
Approximate Block Height (inches)	9.25	9.25	9.25	9.25	9.25
Main Caps	**	**	**	**	†
Tappet Bore Diameter (inch)	.843 to .844	.843 to .844	.843 to .844	.843 to .844	.843 to .844
Cam Centerline Height (inches)	7.44††	7.44 ††	7.44 ††	7.44 ††	7.44 ††

* 2.5585 to 2.5595 inches
** Cross-bolted with two vertical bolts
† Cross-bolted with four vertical bolts on three center mains
†† Next tallest cam height is 6.125 inches on Chrysler small-blocks

four round plugs contain the four MDS solenoids if they are used.

The extra-high cam position is 1 inch higher than on previous V-8 engines. The hollow dowels that locate the front cover, one per side, are visible in the middle of each front cover sealing surface. Note the two short grooves in the cam bearing shell (top). The block casting number is to the right.

The top of the block, or the tappet chamber, is not open but is completely covered. The head's deck surface is extended upward to meet the tappet chamber cover to seal the tappet chamber. Five bolts that are added above the five head bolts toward the cylinder bores help seal the head. The

The close-up of the number-1 main cap shows the main oil feed hole in the center with its unique slot (groove) that tapers at about 4:00 o'clock toward the right and extends to the left about an inch. This matches up to a unique two-hole main bearing shell.

the Mopar big-blocks. That's where the similarities cease and unique aspects begin to stand out. It has a skirted-block similar to the Gen II (426) Hemi and cross-bolted mains like the earlier Hemi, but the Gen III Hemi cross-bolts all five mains rather than just the center three as done on the Gen II 426.

Another unique feature is the very high cam location relative to the crank centerline. It is more than an inch higher than the previous Mopar/ Chrysler V-8 engines, wedges, or the two previous Hemi generations. Next is the very large diameter of the cam bearings, especially the number-1 bearing. If you look at a Gen III bare block, what may strike you first is that there is no front chinawall and it has been replaced by a high-point extension of the two deck surfaces, making the front face into an "A" shape.

Any max-performance engine project begins with an actual cast-iron or aluminum block. If it's new, it's warranteed and it's a clean slate for building an engine. If it's used, you need to inspect it to be sure that it's a good platform for building an engine. In the 1960s, buying a new car with a Gen II (426) Hemi in it and removing the engine for your max-performance project may have been a way to obtain the hardware required because the cars often cost about $3,500 to $4,500 new. Today, the new Gen III Hemi vehicle costs more than $25,000 so this isn't perhaps your first choice. Crate engines are also another source of hardware.

In the 1970s, no one allowed a 426 Hemi engine to go to a salvage yard. They kept it at home and sold it or built it into another car. New engines last much longer but with 14 years of production, mileage may be high on some. Although accidents still put Hemi III engines into the salvage yards, the yards considers them "new" engines for more expensive cars, so the salvage costs are higher.

Remember that in 1964, this basic situation (Hemi blocks parts availability) created Chrysler Performance Parts, which became Direct Connection, which evolved into Mopar Performance Parts. Sourcing a bare Gen III Hemi block from Mopar Performance Parts is also possible today.

The front face of the block includes the large round hole for the variable valve timing (VVT) solenoid used on newer VVT blocks (2009 and up), in center, just above the large cam bearing. The solenoid is held in place by one bolt.

VVT

VVT has been available since 2009 but not 100 percent; most 6.4's front above mains moved forward about .600 to line up with the new front cover. The cam is also about .550-inch longer and about .180-inch larger in diameter. All those changes are required for the additional oil passages that are needed to operate the cam phaser. The phaser changes the installed cam centerline. It closes the intake valve relative to bottom dead center (BDC) and it opens the exhaust valve closer to BDC, which are the key events.

Bores and Bore Centers

The Gen III Hemi has a wide range of cylinder bores that have

It is difficult to measure the block's bore centers directly. However, you can measure the cylinder bore, shown here as 4.09, the standard for the 6.4/392 Gen III Hemi blocks.

Once the exact bore size is known, you measure the distance between the two adjacent bores (.370 inch in this case) and then add the bore size (4.09) to equal the block's bore center measurement (4.46 inches). This thickness becomes important as the bore size increases and moves closer to 4.185 inches.

been used in the various production engines from 3.917 to 4.09 inches in the cast-iron engines and up to 4.125 inches in the aluminum block with even larger bores are possible in the aluminum block. All of these Gen III Hemi blocks use 4.46 inches for the basic bore center, the distance from the center of one bore to the center of the adjacent cylinder. This 4.46 bore center is the same as on Mopar small-blocks and both the A-engine and newer Magnum versions.

Cylinder Block Casting Numbers		
Year	Engine	Block Casting Number
2009	5.7 Eagle Gen III Hemi	53021314DR
2011	6.4 Apache SRT Gen III Hemi	05037471BF
2014–2017	6.4 Big Gas Gen III Hemi	05044547AD
2011–2017	Aluminum blocks	Several bore sizes
The Big Gas block is the best cast-iron Gen III block to date. It has bigger bores, shorter water jackets, and lots of nice stuff.		

This head deck surface shows the head's six-bolt head attaching pattern. Five large head bolts are in the bottom row, five large head bolts in the middle row, with a hollow dowel on each end of the row, and five smaller bolts across the top row. Each cylinder has two large bolts around the bore at the bottom, two from the middle row, and two small ones from the top row; six total per bore.

The block's casting number can be used for identification. It is located on the driver's side front wall or the front face of the number-1 cylinder. It is below the deck surface and above the front cover dowel pin.

Displacement

The largest of the current production Gen III Hemi is the 6.4, or 392, engine. Mopar sells a Drag Pak engine based on the aluminum block that has 426 ci using a 4.125-inch bore and a 4-inch stroke, also available from Arrow Racing Engines. Engine builders are working on a 440 version and even larger ones are in the works.

Head Bolt Pattern

Gen III Hemi engines also use a six-bolt pattern around each cylinder. Four large bolts are around each chamber and then two additional small bolts. These two smaller bolts are in a line directly above the large bolts because the block extends upward to cover the tappet chamber and this surface provides anchors for the bolts and increases block stiffness.

Identification

If you are used to looking at bare Gen II Hemi blocks, your first thought when looking at a bare Gen III Hemi block is that it looks small. Based on the block's bore centers, it is actually a small-block in size. In its

Cylinder Block Casting Numbers		
Year	Engine	Casting Number
2003–2008	5.7 Hemi	53021319AG/CB
2009–2015	5.7 Hemi	53021319DL/DK*
2006–2011	6.1 Hemi	NA
2011–2015	6.4 Hemi	5037473BE
2014–2017	6.4 Hemi Big Gas	"BG"**
2011–2016	426 Aluminum	P5153896AA

* The "DK" casting is the VVT version of the 5.7 block.
** The 6.4 Big Gas engine is the name attached to the heavy-duty truck 6.4 engine, which uses the same full casting number as the 6.4 passenger car engine and adds a large "BG" inside the bellhousing area.

This chart is only a general guide. You may find other numbers because Fiat Chrysler Automobiles (FCA)/Chrysler often changes the last two characters, such as "AA" in the aluminum block number. As the engine evolved, the "AG" became "CB," but not every letter set ended up being cast into the block. Another change is that these engines are still evolving through the production years with additional vehicle models.

The Ram truck 6.4 Gen III engine is known as the Big Gas engine and block. It was introduced in 2014 and was rated at 410 hp and 429 ft-lbs of torque. The engine's compression ratio was dropped slightly to 10:1. The key identifier for this engine is the "BG" 6.4 cast-iron block. As the best performance cast-iron block to date, it features a shortened water jacket, which makes it much stiffer and stronger. Note the short 2-inch vertical ribs that help stiffen the block. It also has thicker cylinder bore walls in the major and minor thrust directions.

Block Height Calculation

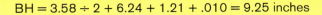

The block height (or deck height) is defined as the distance from the center of the crankshaft to the top of the block's deck surface, measured along the cylinder bore centerline. The Gen III Hemi's production block height is 9.25 inches. You don't always know a block's history; it may have been milled at a previous rebuild. However, you can calculate its height using the following equation:

$$BH = S \div 2 + RL + CH + DH$$

Where:

BH = block height

S = stroke

RL = rod length, center-to-center

CH = compression height of piston

DH = deck height of piston, measured in the actual block

The tricky component is the piston's deck height because it sounds similar to the block height definition, but it is defined as the distance from the top (flat) of the piston at top dead center (TDC) to the top of the block's deck surface. Typically, it's measured with a dial indicator or a bridge, which includes a dial indicator.

If you have a dished or domed piston, the actual top of the piston is the flat part at the outside edge that is not part of the dome or dish. On the Gen III Hemi engines, the piston tends to *not* be flat, but it looks flat; stay at the outside edge to be sure.

Here's an example: On a 5.7 Gen III engine, the stroke is 3.58 inches, the rod length is 6.24 inches, the stock piston compression height is 1.21 inches, and the piston deck height is .010 inch below the deck.

$$BH = 3.58 \div 2 + 6.24 + 1.21 + .010 = 9.25 \text{ inches}$$

If you measured this engine's piston deck height at .0 inch or .010 above the deck, you know that the block has been decked .010 or .020 inch or that the pistons have been replaced. ■

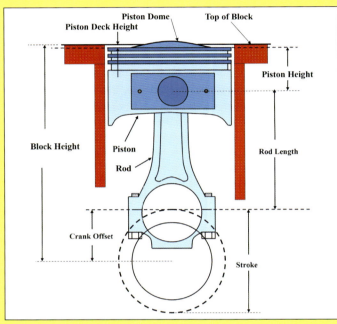

The block height calculation helps determine if the block has been milled during a rebuild or if it has been rebuilt previously. These numbers for the various heights are key to the compression ratio numbers and the various requirements for the use of stroker cranks.

Actual crate engines continually evolve and this is one of Mopar Performance's latest crate engines with 385 hp and 485 hp. MP also offers some accessories that can be quite helpful in any Gen III engine swap.

usual position, the block's pan rails sit square to the floor because of the skirted design. The most unique feature of the V-8 bare block is the high point made by the front face of the bare block. The front face is in the shape of a large capital "A"; it's lovingly called the "bird house."

The basic bare Gen III Hemi cast-iron block weighs about 195 pounds. The aluminum block weighs about half that, about 100 pounds, slightly more for small-bore versions and slightly less for big-bore versions.

In the usual block position, the most notable aspect of the aluminum block is its sleeves. They allow the aluminum block to service many different bore sizes or overbores unlike cast-iron blocks, which are typically limited to .020 to .030 inch over the production bore size.

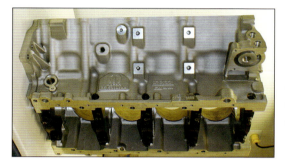

The aluminum block is designed to replace or service all production engines so it has the stock crank position sensor location (left rear above pan rail), the stock four-bolt motor mount bosses (center toward top), and the stock oil filter location (right).

Several aftermarket companies offer various rotating assemblies. This is Eagle's complete stroker package that is available up to 4.050 inches, and it has matching rods and pistons. It is offered for both the 5.7 and 6.4 basic blocks.

Rotating Assemblies and Long-Blocks

When you are building a relatively new engine, sourcing basic hardware can be time consuming. Salvage yards want top dollar for Gen III Hemis because they have low mileage and are new. If you have a basic block, you might consider using a rotating assembly that includes crank, rods, and pistons that are all matched. These are available from Eagle, Indy Heads, Scat, Manley, and Callies/Compstar.

Another approach is a long-block from a Chrysler dealer. The long-block is an assembly that does *not* include valvecovers, intake manifold, front cover, or oil pan. You might try a 5.7 Eagle (68259163AA) or a 6.1 (68253461AA) as a starting point, available at any Chrysler dealer.

Aluminum Block

The basic aluminum Gen III Hemi block is kind of a jack-of-all-trades. It is the best block to use in any racing class where it is allowed. The cast-iron blocks are thin-wall cast blocks and as

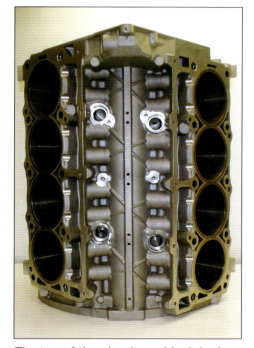

The top of the aluminum block looks similar to that of production cast-iron blocks with the four holes for the MDS solenoids, shallow tappets and sealed tappet chamber, and the six-bolt head attaching configuration.

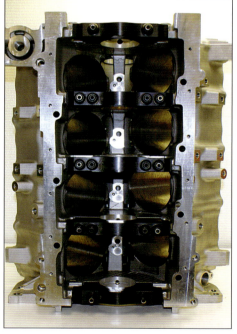

The most noticeable aspect of the bottom of the aluminum block is that it uses four vertical bolts in the center three main caps. Here, you can also see the machining for the pin oilers; in the center, between each set of main caps.

Aluminum Block Features

- Lightweight aluminum construction offers a 100-pound weight reduction over Hemi Gen III production cast-iron blocks
- Full-skirted block design, similar to the 426 Hemi for increased strength and rigidity
- Cross-bolted main caps for added strength
- Water jackets redesigned to accommodate siamese bores
- 8620 HR billet steel main caps with four-bolt center caps, plus cross-bolts on each main
- Camshaft bearing support area increased to permit the use of 60-mm roller cam bearings, if desired
- Built with stock-size lifter bores, which can be opened up to 1.060 inches to allow the use of bushed-style lifters
- Crankcase clearance designed for a 4.125-inch-stroke crankshaft; a 4.250-inch stroke fits with machining around the oil drainbacks
- Capable of supporting a dry-sump package for reduced windage
- Additional boss in the front timing chain area for an idler (double roller chain)
- Additional boss for a front gear drive, plus attachments for a front engine plate
- All Hemi Gen III dress items (front cover, oil pan, oil filter mount, and front-end accessory drive) fit this aluminum block

such should be overbored by .010- or .020-inch max. With aluminum blocks, the bores can measure 4.155 inches, with 4.185 inches possible in the near future. Performance head gaskets are needed to seal this bore size.

This aluminum block front wall shows the large number-1 journal used in the Gen III engines and the high location of the cam. Note that there is no VVT hole above the cam bearing. The aluminum block is designed as a replacement for the 2003–2008 5.7 and 6.1 non-VVT blocks. VVT blocks have the top part of the front wall moved forward about .500 inch.

The casting number for the aluminum block is located above the passenger-side pan rail, ahead of the crank sensor hole.

The passenger's side of the Gen III aluminum block has four or the five cross-bolts; the pan rail blocks number-5. The oil filter fits at left. The crank sensor is at the right. The main caps are made of high-strength steel.

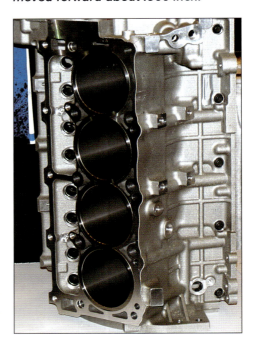

A second aluminum block is made by Gen3 Performance Products and seems to be designed mainly for all-out racing rather than street or dual purpose use. It has the production motor mounts, so it could bolt into a production chassis engine compartment. It has sleeves that can allow 4.185-inch bores and can use a 4.25-inch crank (468 inches). It has more external ribbing than the other aluminum block, which adds rigidity. Typical of parts designed for racing, there are many options or special features that can be added as desired.

The standard bore size on these aluminum blocks is 4.125 inches. It is the basic bore size for the 426 crate engine and the 426 Drag Pak engine. However, many other bore sizes are available, such as a 4.055-inch bore (6.1 style).

Semi-finished bores are available as well a builder's special.

A great advantage of the aluminum block is that it can be repaired. If you wear out or damage a cylinder, you can simply replace a sleeve or go to a larger bore size.

Through 2016, the supercharged engine used in the NHRA Drag Pak 354 engines had to use a cast-iron block. Availability of these blocks in good condition has become an issue for NHRA racers. The 426 naturally aspirated version had no problems, so for 2017, the aluminum block is allowed to be used in the 354 small-bore configuration.

Engine Details

The following discussion includes specific hardware, technical terms, and the advantages of a max-performance engine program.

Block Height

The Gen III Hemi block height is quite short at 9.25 inches. Earlier Mopar small-blocks were about 9.60 inches and the Gen II Hemi was based on a big-block and had a 10.72-inch deck height. This shorter height makes the pistons and rods shorter and much lighter and brings the heads closer together.

Chinawall

Perhaps the most obvious visible feature of the Gen III Hemi block family is the high point (birdhouse) in the front face of the block. This affec-

There is no chinawall on Gen III Hemi blocks. The two deck surfaces are extended upward at the stock 45-degree angle until they meet at the "birdhouse." This feature makes Gen III blocks look different from other V-8s. Halfway between the cam bearing hole and the point is the VVT solenoid hole. This is only used on the 2009 and newer blocks.

tively eliminates the front chinawall. Virtually all previous V-8 engines have a chinawall, which is created by connecting the two 45-degree deck surfaces with a horizontal wall that also helps seal the tappet chamber.

On the Gen III Hemi block, the deck surfaces are extended upward and toward the center at their existing 45-degree angle until they meet in the center, forming a large "A" shape. The left surface of the A-shape is the left deck surface and the right surface is the right deck; no chinawall.

Mains

The Gen III Hemi uses five main bearing bulkheads, as do other modern V-8 engines and similar to Gen II Hemi engines. The production Gen III Hemi uses two vertical main cap bolts and smaller cross-bolts on all five main caps while the production Gen II Hemi cross-bolted the center three main caps only.

The aluminum block version uses four vertical main cap screws. Modern Muscle can also add this four-bolt feature to the Big Gas 392 block. The unique aspect of this block is the center main (number-3),

It is easy to see the five main caps and the two vertical bolts in each cap. These five caps are cross-bolted. (Three can be seen here; the pan rail hides the other two.) All five mains being cross-bolted is a unique feature of Gen III production engines. The Gen II 426 Hemi was only cross-bolted on the center three mains. The capability for all five was added to the Race Gen II versions in the 1990s.

The typical cross-bolted main cap has two large vertical bolts and two smaller side bolts (not visible here). This block is upside-down, and the forward arrow is on the right number-3 cap only. The others are on the left. Note the unique machining for the number-3 thrust bearing below the parting line on this number-3 main bulkhead. It uses only a half-moon insert for a thrust bearing with one per side.

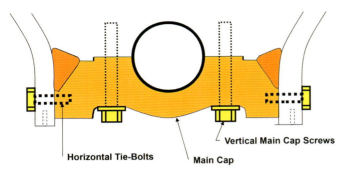

Gen III Hemi Main Bearing Cap

All Gen III blocks use cross-bolted mains on all five main caps. Gen II engines only used them on the center three. The aluminum block (Arrow Racing engines) uses four vertical bolts by using a special cap and machining that allows the extra vertical bolt between the standard vertical bolt and the smaller cross-bolt.

Several bolts holes are in the side of the Gen III blocks, but the four motor mount bolt bosses form a square that is slightly to the right of center and starts at about a 1.5 inches below (above in photo) the deck surface.

which takes the thrust. The number-3 main bearing shell is the same as the number-4 and the number-2; no flange. A half-round insert on the front and rear of the number-3 bulk-head/main cap takes the engine's thrust. The recess that accepts this insert is machined into the block not the cap. The crank and the cap hold them in place. The inserts look like half-round hockey sticks, flat in one plane and curved in another.

Block Oiling

The oil pump on the Gen III Hemi block is located at the front, inside the front cover. The oil filter is located at the front passenger's side on the block. It is similar to the Gen II Hemi except that the filter is on the opposite side. That is about the only similarity that the new oiling system has with previous production engines, big- or small-blocks. From the oil pump/filter oil return to the block, the oil travels

to the main oil galley, which feeds the crank's mains and the cam journals. Then it passes through passages in the block and head to the valvetrain and into the rocker shafts, one intake shaft, and one exhaust shaft per head. From the shaft, the oil goes to the rocker arms, which delivers it to the valves, springs, etc. The rocker arms deliver the oil to the pushrods. The pushrods deliver the oil down to the hydraulic lifters (internal).

Motor Mount Bosses

The engine's motor mounts bolt into each side of the block in the approximate middle of the side of the cylinder bank. This system is similar to the three-bolt system used on the Magnum (5.2 and 5.9) small-block engines except that the Gen III Hemi uses a four-bolt system on each side. The Gen II Hemi used a three-bolt system, but it was located at the front of the block and lower, closer to the pan rail.

Schumacher Creative Services and others offer motor mounts for swapping the Gen III Hemi into earlier models. The four-bolt, mid-side location allows a lot of flexibility in mount fabrication for custom vehicles.

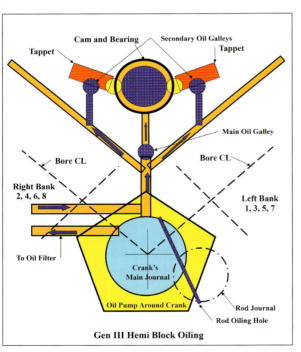

Gen III Hemi Block Oiling

It is very difficult to show the oiling paths in a block. The main oil galley is located about halfway between the main journal and the cam journal, in the center of the block. Not shown here is that the inside of the hydraulic tappet is oiled from the valve gear. The valvetrain (heads) is oiled through the block and heads similar to Gen II engines.

TDC Boss

The TDC boss and sensor is located on the passenger's side of the block toward the rear. It consists of a machined hole for the TDC sensor and a threaded hole for the attaching

The TDC sensor boss is located at the rear passenger's side of the block, by the number-8 cylinder in the outside wall. It is about 3 inches above the pan rail, and aligned with the crank sensor wheel (gear-teeth below pan rail toward bottom left), which is just left of the number-8 crank counter-weight. It is held in by one bolt. The main cap cross-bolts are visible here, just above the pan rail.

This closed-tappet chamber between the two cylinder banks shows two of the four MDS locations with plugs. The tappet chamber is very shallow (almost flat) because the tappets are at a very shallow angle, 15 degrees above horizontal, or 75 degrees off vertical. The cam being raised so high above the crank in the block allows for a shallow cam angle.

bolt. It somewhat lines up with the number-8 cylinder. It is a small hole, but it is very important. If you use MPI (multi-point injection) as the Gen III Hemi does (production and race) this TDC sensor tells the engine control module (ECM), or computer, where TDC is. A timing wheel attaches to the crank, which the sensor looks at and can tell precisely where the crank is in its rotation at all times. Even if you want to run a distributor and carburetor, you might want to keep the TDC sensor (see Chapters 9 and 10 for more details).

Tappet Chamber

In a typical V-8 pushrod engine, the intake manifold or a gasket and shield seals the tappet chamber. This configuration allows you access to the tappets if the intake manifold is removed. With the Gen III Hemi, the block seals the tappet chamber, and removing the intake does not allow you to access the tappets. The heads must be removed first. This keeps hot engine oil off the bottom of the intake runners in the manifold. This chamber sealing is

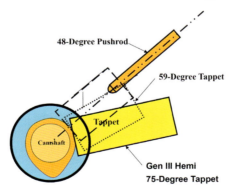

Small-block racing blocks used a 48-degree tappet angle from vertical and the Gen II Hemi was at 45 degrees while the production small-blocks and the Gen I blocks were at 59 degrees. Gen III blocks are at 75 degrees, almost flat, or horizontal.

accomplished by extending the two deck surfaces upward to enclose the tappets. This means that the head must be extended upward to meet the block, which means the tappets and pushrods are now inside the sealed block and head. The tappets *must* be installed *before* the heads are installed. The net result of the wider deck surface is that the block and head are much stiffer and stronger.

Tappet Angle

The tappet angle is not easily changed. Gen II Hemi blocks (and Mopar big-blocks) were at 45 degrees. The production small-blocks and the Gen I Hemi were at 59 degrees. The Gen III Hemi is at approximately 75 degrees off vertical, or 15 degrees off horizontal. This angle change is caused by the height of the camshaft, which was raised more than 1 inch from previous heights (Mopar small-blocks).

Packaging

The packaging is not something that the average performance enthusiast considers. However, it is very

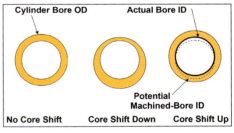

Today, block castings are machined on CNC computer-controlled machines; this precision machining process centers the actual machining on the cylinder bore (right) and centers the wall thickness. This centering process reduces or cancels the affects of any core shift. Centered bores (left) mean no core shift. A core-shift problem (center) could be in any direction: left, right, up, or down.

The block has cylinder head drainbacks. Symmetry would suggest five, similar to the number of big head bolts per row. The oil drainback is the hole next to the head bolt, which is lower (farther away from the deck centerline). Cylinder number-2 (far right) has no drainback, only a head bolt.

Gen III Hemis locate the cylinder heads on the block using two dowel pins similar to the other Mopar V-8s. The other Mopar V-8s use two solid dowel pins; the Gen III uses two hollow dowels. The dowels are located around head bolts number-8 and -10 in the torque sequence or the bolt to the top left and the bolt to the top right on each deck surface. Each dowel pin has a height of about .486 inch with an OD of .630 inch and an ID of .505 inch.

important to Chrysler engineers who design cars. Will the engine fit in the engine compartment? One of the two key elements in the design of the Gen II 426 Hemi cylinder head was the distance between the shock towers on the production B-Body (Belvedere/Coronet) vehicles. That is part of packaging.

With the 4.46-inch bore centers, the front-to-rear length is similar to that of a small-block. The short deck height makes the engine narrower and shorter. The cylinder head uses a shallow chamber and more vertical valve angles, which helps narrow the cylinder head.

The net result is that the Gen III Hemi assembly is smaller than many small-blocks and may fit in many places that other Hemi engines do not, or at least without major modifications. It's the perfect choice for many street rods and street machines.

Cam Height

The camshaft in Gen III Hemi engines is raised a large amount (more than 1 inch) and is very obvious visually. Most production Chrysler V-8 engines have used a cam height of around 6.125 (small-block) inches with all big-blocks being less. The Gen III Hemi uses a cam height of 7.44 inches. That's about 1¼ inch higher. One advantage of this configuration is that the cam is a long way from the crank and rods. For max performance, this means longer-stoke cranks can fit more easily. In addition, the pushrods are shorter, which makes them lighter and stiffer at the same time.

Bellhousing Pattern

The Gen III Hemi uses the Mopar small-block bolt pattern, which is the same as all A-engine (1964–1993) and all Magnum (5.2 and 5.9) engines (1992–2003). This feature is helpful in sourcing bellhousings, safety bellhousings, and fitting transmissions to the engine. This feature can be helpful when engine swapping.

Oil Returns

The Gen II 426 Hemi used two oil returns from the cylinder head, one on each end of the head on the exhaust side next to the valvecover rail. The Gen III Hemi uses four oil returns. One is located between each

Although not used in all Gen III blocks, many engines have piston oilers or sprayers that are difficult to see even if the block is disassembled. In the center, you can see the small spray bar with a sprayer pointed at each cylinder on a journal, four sprayers per engine. They attach directly to the bottom of the main oil feed.

Almost all production V-8 engines use a two-piece rear seal on the crank. Gen III Hemis use a 360-degree one-piece rear seal. Five bolts hold the aluminum retainer to the rear of the block.

cylinder, on the exhaust side next to the valvecover rail and one on the end of the head. However, the front drain hole is omitted on each head, which makes for a left and a right head; they are not the same in production. The block is machined to match these oil drainbacks in the head. If an aftermarket head is used, the front drain should be plugged because there is no matching drain in the block.

Crate Engines

A crate engine is a fully assembly engine minus some accessories; it is a convenient way to source all the parts that you may need for a max-performance Gen III Hemi engine project such as a street rod, a modified street machine, a show car, or any other custom application. If you are maxing out a production vehicle, you already have the basics and don't need a complete assembly, unless the current assembly is broken.

Gen III Hemi crate engines come in the 5.7, 6.4, and 426 aluminum block.

Arrow Racing Engines offers a 362-ci "built" engine designed for specific circle-track racing series in Canada. It makes 530 hp with a 390-cfm Holley 4-barrel.

Indy Heads offers Gen III Hemi crate engines. The company currently has a 400-hp 6.1 with a 435-hp CNC-ported head option and a 600-hp 426-inch upgrade, based on the 6.1 engine with a 4.09-inch bore and a 4.05-inch stroke. Both are based on the cast-iron block and feature an 8-barrel in-line intake with two 500-cfm Edelbrock carburetors.

Indy Heads also offers several options on the various Gen III Hemi hardware from CNC-ported 5.7 and 6.1 heads to fully CNC-ported 6.4 heads (more flow) and many manifold and induction choices, including 4-, 6-, and 8-barrel carburetor options to MPI options.

Stanton Racing Engines, Modern Muscle, Hughes Engines, and Bischoff Engine Services (BES) also offer crates many options.

Racing Crate Engines

Various Gen III Hemi engines have been competing in several NHRA Stock and Super Stock drag racing classes for the past seven-plus years and are very competitive in these classes. The Drag Pak Challengers is a popular package and two or three engines options are offered per year. These special-built cars were not allowed to have a supercharger until the Hellcat went into production. In production, the Hellcat engine is built as a 6.2 engine, or about 378 ci.

To run in the NHRA classes, it must be built at 354 inches. The first versions were based on the cast-iron block, but for the 2017 racing season, the optional aluminum block can be used because it was already legal in the naturally aspirated version. Even with the aluminum block, the engine must still be built at 354 inches.

Unlike the Drag Pak package that was designed for drag racing, the basic Gen III Hemis are not installed in lightweight production vehicles and therefore do not race in many categories outside of drag racing. Participation of this engine family in circle-track racing has been limited. The Canadian Pinty's racing series competes on both circle track and road racing courses.

Racers have been pleased with the new 362-inch built engine based on the Gen III Hemi and the

Crate Engine Specifications				
Year	Engine	Horsepower	Torque (ft-lbs)	Part Number
2016–2017	5.7	383	417	68303088AA
2016-20-17	6.4	485	475	68303090AA
2015	426	*	*	*
* MP offered the 426 aluminum crate engine, but intake availability caused supply problems and it is not currently offered. However, Arrow Racing Engines still offers aluminum 426 crate engines in several variations with the max-power package that is about 780 hp. The large cam and tall intake used on this version may not be suitable for all applications.				
The 5.7 crate engine rated at 383 hp replaces earlier 5.7 crate engines that were rated around 370 hp. The 6.4 crate engine rated at 485 hp replaces an earlier 6.1 crate engine that was rated at 425 hp and earlier versions of the 6.4 engine.				

For the 2016 NHRA racing season, this 354 supercharged crate engine was offered for use in the NHRA Challenger Drag Pack cars. It uses a Whipple supercharger and produces 354 ci, both required by the NHRA for the racing class. Note the wide supercharger drive belt (right) and the Moroso rear-sump oil pan (bottom). Original engines used cast-iron blocks, but for 2017 the aluminum block is legal; It is available from Arrow Racing Engines.

aluminum block. Arrow Racing Engines has a complete engine for this series, which produces 530 hp with the required 390-cfm carb.

Performance Packages

To get the most out of any engine package, the various parts must work together. The Gen III engine family does not fit into the typical "horsepower" categories (400, 500, etc.) because the stock 6.4 makes 100 hp more than a stock 5.7 for only a 45-ci difference. Therefore I have used several methods to group the parts based on the method that seemed to work best for that specific category.

Cylinder Block Overbore Specifications				
The best block for an application is often based on the bore size required for a particular application.				
Basic Engine	Block	Max Rebuild (inches)*	Bigger Bore**	Racing***
5.7	Cast Iron	.020 to .030	6.4 and Aluminum††	BG† and Aluminum††
6.1	Cast Iron	.020 to .030	6.4 and Aluminum††	BG† and Aluminum††
6.4/392	Cast Iron	.020 to .030	BG† and Aluminum††	BG† and Aluminum††
426	Aluminum	4.15	4.185	4.185
* Standard block safe overbore. ** If bore's bigger than .020 to .030 inch, oversize are required. *** Aluminum block has many sleeves; the BG block would have to be sleeved down to size. † BG stands for Big Gas 6.4 Truck version. †† Aluminum block has several options for sleeves, which allow for different bore sizes.				
The BG version of the 392 block may allow more over-bore size, but it is too new to be fully evaluated.				

Machining Operations

As soon as you have selected a block for your project, you should begin performing the required machining operations. The Gen III Hemi is a thin-wall casting and, therefore, should not be machined heavily. Overbores should be limited. I recommend ordering an aluminum block in the configuration that you desire to save time and hassle as well as minimize the amount of machining.

Bore and Hone

Cast-iron Gen III Hemi blocks should only be bored to .010- or .020-inch oversize. These are thin-wall castings and by overboring them, you make the bore walls too thin for max-power applications. If you are planning a max-performance engine, you should use the block in its "stock" bore configuration to maintain maximum strength. Do not try to bore out a 5.7 block to a 6.1 (4.055 inches) or 6.4 bore size (4.09 inches). In addition, I don't recommend boring out the 6.1 block

to the 6.4 bore size. If you want to use the 4.09 bore size, begin with a 6.4 block or the aluminum block. Remember to always hone the block after the boring operation using a honing plate.

Honing plates for the Gen III Hemi are not readily available because of the unique six-bolt attaching pattern. A honing plate is a 1- to 2-inch-thick steel or cast-iron flat plate with all of the head bolt holes and the cylinder bores (slightly larger than the max overbore size) machined in. It looks like a very thick head gasket. It is also quite heavy, so handle with care.

It is attached to the cylinder block with the same size of head bolts as the actual head and torqued to the same specifications. A honing plate is designed to simulate the stresses, distortions, and wall movements that are usually caused by the installed and fully torqued cylinder head. It is generally recommended that the main caps be installed and properly torqued during the honing operation.

Decking

As a general recommendation, you should keep any deck milling to a minimum, perhaps .010 or .020 inch max. The main reason for this limit is that the piston is close to the head and milling decreases the piston-to-head clearance below accepted limits. This situation would cause you to clearance the piston to gain head clearance so you increase the amount of work required (extra machining on the pistons) and probably your costs also. Change the piston to gain compression ratio.

The previous V-8 aluminum blocks typically required the sleeves to protrude above the deck surface by a few thousandths of an inch. This relationship is not required on the Gen III Hemi aluminum block; they can be at the same height or flat to the deck.

Sleeving

In many cases, you need to sleeve cast-iron blocks to repair them. Once the Gen III block has been cleaned up, your machine shop may tell you that one or two of the block's cylinder bores are damaged and need to be fixed. Basic wear and/or scratches in the bore are the most common causes. No matter the reason, sleeving by a machine shop (or resleeving on an aluminum block) should take care of it. If the process is executed properly, the rebuilt engine should be as good as new.

Sleeving is an option for changing the bore size of cast-iron blocks. There are no 5.7 race blocks and the smallest 6.4 race block bore is 4.09 inches, which is the Big Gas block. Perhaps your current race engine is worn out. If your last rebuild was at the maximum bore size for the block, it may be time for another rebuild or you may need to scrap the block. Sleeving all eight cylinders is one solution.

So, what if you want a race block with a bore of 3.97 inches (a .060-inch oversize 5.7)? With small bores or worn bores, sleeving is a reasonable approach to the desired end. Try boring the block in the cylinder sequence of (left bank) number-3, -7, -1, and -5. For example, if you

want to build a 325-ci Gen III Hemi with the readily available 3.58-inch stroke based on the 6.4 block (original 4.09-inch bores), you need a 3.80-inch bore. Sleeving is the only way to get to this bore size.

Sleeving is also an option to repair an aluminum block. Alumi-

You can resleeve an aluminum block for a bore size change. Similar to a repair, the eight sleeves in the block are replaced with new ones. Sleeves must be replaced using heat, so you should return the block to the manufacturer for this operation. Sleeves are made by Darton Sleeves. This example from Arrow Racing Engines is for an aluminum block, which can have bores up to 4.15 to 4.185 inches.

num blocks already have a sleeve so if it is damaged, the sleeve is removed and a new sleeve is inserted. This process uses heat, so it is best left to the manufacturer or a professional engine shop.

Align-Boring

Align-boring is an operation that may not be required on the Gen III Hemi blocks. It must be performed by a machine shop. However, because most Gen III blocks are new, the main bearing bore alignment should still be fine. The exception might be a block that has had a catastrophic failure and has been welded for repair.

Cylinder Loading

Gen III cylinder bores are round, but the cylinder walls inside the water jacket generally are not because of the water jacket. In a typical engine with a full water jacket (not a siamesed bore), the actual cylinder's bore-wall thickness varies; it tends to be thin toward the front and rear of the block and thick toward the top and bottom.

This double egg shape is basically by design and is the desired shape (better than round). The reason is that the cylinder walls in each cylinder bore are not loaded evenly. The piston skirt transfers the piston loads to the cylinder wall and because the piston pivots about the piston pin, the piston skirt does not load the wall equally around the 360-degree bore. There is very little load toward the front of the engine and very little toward the rear.

The engine's direction of rotation causes one side of the bore wall to load higher than the other. The highly loaded wall is called the major thrust direction and the other side is called the minor thrust direction. With the engine's normal rotation (clockwise), the major thrust direction is toward the passenger's side of the engine, which means on the left bank, or 1-3-5-7, the major thrust side is toward the tappet chamber and on the right bank, or 2-4-6-8, the major thrust side is toward the outside of the block. You want more bore-wall thickness in the major thrust direction.

This double-egg-shaped cylinder wall is used on BG 6.4 blocks that allow bigger bores or more overbore. The water jacket is also shorter, which makes the BG block and cylinders stiffer and stronger. ∎

Max-Performance Technology

In many cases, racing helps find solutions to problems, which provides technology that can improve overall performance. In the late 1960s and early 1970s, the 426 Hemi Gen II helped solve some problems, which helped improve production engines. In the mid-1980s, production technology helped improve racing engines. By the late 1990s and the early 2000s, earlier race technology was everyday knowledge and the production engines of the early 2000s reaped the rewards of this technology boom. The Gen III Hemi was designed and produced based on much of this technology.

Stress-Relieving

Today stress-relieving of a block is not required. The exception is a block that suffered a catastrophic failure. Because the engine is still in production, a new block is a much better choice.

Sonic Testing

Sonic testing was originally used as a selection process to decide which one of your blocks was the best to build into a race engine. If you don't have several blocks, you should use it to determine the amount of overbore. For sonic testing to be useful, it must be done before the block is overbored.

Cam Bearings

The cam bearings in the Gen III Hemi blocks are larger than the just-under-2-inch bearings that have been used in Chrysler small- and big-block V-8s for many years going back to the 1950s and early 1960s. Both are considered "sliders."

In the 1990s, roller cam bearings in 50-mm designs became popular in certain racing categories, and even-

tually they grew into the 60-mm roller cam bearings and 60-mm cam designs. Today, the aluminum block has the provision to allow for the addition of 60-mm cam bearings, but production-based cast-iron blocks do not allow for this modification.

Head Gaskets

The production 5.7 engine does not use the same head gasket as the 6.4 because the bore on the 6.4 is much larger. Another gasket is used for the 426-ci package (4.125-inch bore). Bigger bores should use bigger-bore head gaskets. You should not use a big-bore gasket on a small-bore engine. Cometic offers several bore sizes.

Block Filler, Race Only

Years ago many racers added block filler such as Hard Block to the engine's water jacket to make it smaller. The theory was that block filler helps stiffen the cylinder walls. The BG block solves this problem by making the water jacket shorter and making the bore walls thicker. I recommend using the BG block instead of block filler.

Rear Seal

The rear seal on the Gen III Hemi is a 360-degree seal. It fits over the rear flange on the crankshaft. It is held in place by an aluminum retainer, somewhat like the aluminum retainer used on the Gen II Hemi. However, the Gen II retainer slipped into place with two side gaskets and was held in place by two vertical screws. The Gen III retainer is held in place by five screws that bolt into the rear face of the block.

Front Cover

There is one aluminum front cover for standard 5.7 engines and

one for the VVT versions that were introduced in 2009. Some trucks' front covers might be unique from the car versions. If your project is upgrading a production-based Gen III Hemi package, I recommend keeping the front cover and the vehicle together. There is no reason to

The Gen III front cover is a little more complicated than the older Gen II engines. It also serves as the back wall for the water pump. The front crank seal is at the bottom. It is an aluminum casting.

Gen III Hemi water pump bolts to the front cover and is unique to the Gen III engine family. Currently Meziere makes an electric racing-style pump with a unique front cover that has a water pumpback kit. It replaces the front cover and provides passages that connect the water flow from the Meziere water pump to the outlets in the block. The Meziere is not a direct replacement for the production water pump and needs this plate to seal everything up.

This is the back of the Meziere water pump that must be used with an electric pump.

This electric Meziere water pump and backing plate is installed onto an engine built for the 2017 Amsoil Engine Masters Challenge. (Photo Courtesy Red Rocker Engine)

The production serpentine belt for the front accessory drive runs from the left to the right and powers the alternator and water pump; it has two idlers. When a supercharger is added (707-hp Hellcat) a completely separate ribbed belt is added behind this belt.

The front accessory drive on the Gen III engines is a serpentine belt drive. March offers a special billet-aluminum pulley set and also offers various pulley ratios.

change the front cover and it doesn't tend to wear out. The exception is the use of a distributor.

Production engines do not have any provision for a distributor or a mechanical fuel pump. The special front cover available from Arrow Racing Engines resolves both of these problems. Aluminum blocks take the standard (non-VVT) front cover used on the 6.1 engine and the early 5.7s.

Fasteners

Metric bolts fasten the block's main bearing caps and the cylinder heads. Aftermarket companies, such as ARP, offer studs to replace all of these bolts. The studs provide more threads into the block, so it allows the load to be spread along more of

the cylinder wall, which therefore distorts less. Typically, the amount of torque used on a bolt or stud is defined by its diameter.

In addition, studs allow you to quickly remove the caps to check or modify the bearings or pull the head to check seats or guides; when racing, this is an important feature. Typical street or dual-purpose engines are assembled and run for many thousands of miles and long times. Because the engines aren't usually disassembled frequently, you don't see the wear and tear on the threads by the bolts.

TDC Sensors

The TDC sensor bolts to the block. One 32-tooth and two 58-tooth timing wheels are available that bolt to the last counterweight on the crankshaft. Each style of timing wheel has a matching TDC sensor. The sensors and timing wheels *must* be matched. Likewise the TDC sensor and wheel *must* be matched to the computer or ECM (see Chapter 3 for more details).

Main Caps

Five main bearing caps are typically installed in V-8 engines, and the Gen II has a skirted-block design that allows all five main bearing caps to be cross-bolted. On the Gen III Hemi blocks, all five mains are numbered,

The aluminum block has the best example of main caps because it shows the two-bolt cross-bolted mains at number-1 and -5 and the four-bolt vertical cross-bolted mains on numbers-2, -3, and -4.

and on the opposite side of the cap from the number, an arrow indicates the direction of installation.

The main cap bolts are torqued first and the side bolts are torqued next. The aluminum block uses main caps with four vertical bolts instead of two (on the center three mains). On all Gen III blocks, although the number-3 main cap is basically the same as the other four caps, the upper part of the number-3 bulkhead is machined uniquely to hold the special thrust inserts, one on each side of the bulkhead. These inserts are held in place by the crank, main cap, and unique design of the insert itself.

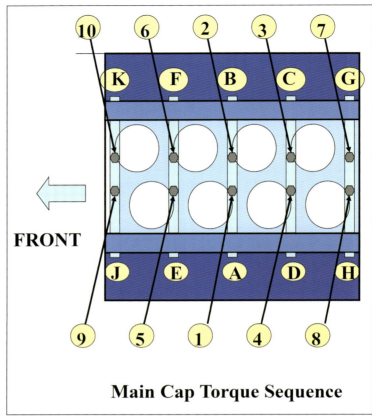

Main Cap Torque Sequence

Gen III main cap bolts are torqued in a two-step sequence, all numbered bolts to 20 ft-lbs and then all numbered bolts plus 90 degrees. After these vertical bolts are torqued, you torque all cross-bolts marked with letters in sequence starting with "A" (to 21 ft-lbs).

This close-up of the four vertical bolt caps used on the aluminum blocks shows the forward arrows on the right side of the top cap and the left side of the bottom cap. Caps numbers-1, -2, -4, and -5 are on one side and number-3 is on the opposite side.

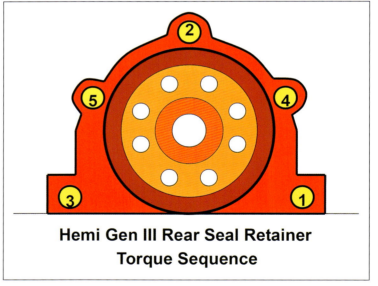

Hemi Gen III Rear Seal Retainer Torque Sequence

The Gen III Hemi rear main seal is held in by five bolts, which should be torqued in sequence to 11 ft-lbs.

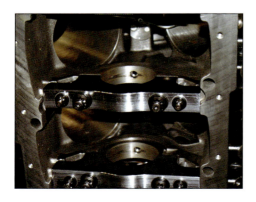

The Big Gas 6.4 block has extra material added for strength, which allows the four vertical bolt-style main caps to be added. Modern Muscle offers this special heavy-duty main cap design. The block must be specially machined to accept these bolts and the cross-bolts, which are similar to those used on aluminum blocks.

CRANKSHAFTS AND CONNECTING RODS

Nodular-iron production cranks are installed in 5.7, 6.1, and 6.4 Gen III Hemi. These stock crankshafts can easily support up to 1.6 to 1.8 hp/ci and are suitable for most high-performance street applications. Although the cranks and rods are fine for naturally aspirated applications, you should upgrade these components with the addition of a supercharger. However, aftermarket forged steel crankshafts from Manley, Callies, Molnar, Eagle, and others are necessary for extreme performance and race applications.

Production cranks have under-cut journals while aftermarket forged cranks use full-radius journals. However, the big news is that the engines do not need to be externally bal-

anced. If you have a used crankshaft, you should take it to a machine shop for a thorough inspection. Assuming that it is still in one piece, a machine shop can usually repair any crank depending upon output levels for the finished engine. Also check the cost of a new, forged crank versus the costs of repairing your existing piece.

The most common problem with cranks, including the Gen III, is that small pieces of dirt may pass through the engine over time (or mileage) and scratch the journals. Your machine shop or crank grinder can re-grind the crank journals undersize and repair this common damage. Typically, the journals are ground undersize by .020 inch. If both the rod journals and the mains are ground

to the same amount undersize, the crank is referred to as a 20-20 crank and the bearings must be matched to these numbers.

Production Cranks

To date, two main stroke lengths have been used for Gen III Hemi production cranks: 3.58 inches for the 5.7 and 6.1 and 3.72 inches for the 6.4. Because the block is shorter than early V-8s and the piston is shorter and lighter, Gen III Hemis do not use any external balance weights currently. All bearing sizes and diameters are the same unlike on earlier

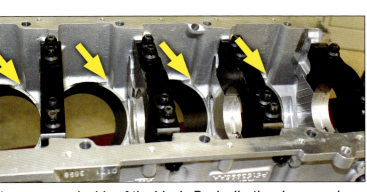

When the 4.00-inch stroker crank is installed in the aluminum block, four smaller reliefs are machined at the bottom of the cylinder bores on each side of the block. Basically, the clearance is ground to the side of the oil drainback boss next to the bottom of the cylinder.

Molnar makes forged cranks that use a 58-tooth crank sensor wheel at the rear. Most Gen III cranks come with this wheel, but there are several versions so use caution.

Mopar small-blocks. All cranks use undercut radii at the edges of the journals, both rods and mains. The aftermarket offers many forged cranks that use a full-radius approach on the journals.

Another unique feature of Gen III Hemi cranks is that they use an eight-bolt crank flange. Although the small-blocks used a six-bolt flange, the Gen III 426 Hemi used an eight-bolt flange to gain added clamping loads for flywheels and flexplates. Many small-block race cranks also used the eight-bolt flange.

With Gen III engines, the unique aspect is the VVT crank. The width of the rod journal on the crank is 1.869 inches.

Multi-Displacement System

Gen III Hemi engines have used the Multi-Displacement System (MDS) since it was introduced in 2004. The system was introduced on Gen III engines and used on about 95 percent of them. The tappet chamber cover has four solenoids that serve as cylinder deactivators.

The MDS basically drops half of the engine's cylinders when you are

Gen III engines use an eight-bolt crank flange similar to Gen II engines and many race small-blocks. Note the small bolt hole at the center top and another on the left side, in the counterweight. Two of the four holes attach the crank sensor wheel.

Common Hemi Crankshaft Specifications					
	5.7	6.1	6.4	6.2	426
Stroke (inches)	3.58	3.58	3.72	3.58	4.00
Main Journal Diameter (inches)	2.559	2.559	2.559	2.559	2.559
Crank Pin Diameter (inches)	2.126	2.126	2.126	2.126	2.126
The 3.58-inch Hemi crank is 1.8 pounds lighter than the 3.58-inch-stroke 360 wedge (small-block) crank.					

cruising and don't need the extra power, which means that you don't use as much gas. How it does this begins with deactivating four of the cylinders, which is what the four solenoids do. Rumor has it that the computer can accomplish this deactivation in 10 milliseconds. The base camshaft (lobe, etc.) is not changed by this system.

It is not used in any of the performance packages, so if you have one, you need to disconnect the four solenoids in the tappet cover.

VVT Crank

The VVT engine was introduced in about 2009. Most 6.4 engines use VVT; the older 5.7 engines did not.

The main difference in the crank is that the number-1 main journal at the front of the crank. The older, non-VVT crank has a tapped area in front

The crank sensor wheel attaches with four bolts to the last counterweight on the crank. Basically, the crank sensor wheel is flat and round with teeth on the outside and a large hole in the center. It has 32 teeth on the driver-side wheel (2003–2008) and 58 teeth on the passenger-side wheel (2009 and newer). See Chapter 10 for more details.

of the number-1 main journal that leads to the snout. The front of this tapered surface locates the crank gear.

On VVT engines, the cam gear is moved forward and therefore the tapered surface on the crank is extended forward by about .46 inch so that the VVT crank gear lines up with the VVT cam gear. The extra width, or thickness, of this tapered section at the front of the crank is easily noticed.

Weight

The Gen III crank is about 2 pounds lighter than older Mopar small-block cranks (360 or 5.9). The standard 5.7 and 6.1 versions of the Gen III Hemi and the small-block 360/5.9 engine use the 3.58-inch stroke, so it is a fair comparison. Long-stroke cranks are heavier.

The front end of this Scat forged crank is extra long, to the left of the number-1 main journal. The VVT engines use an extra length ahead of the number-1 main for oil pump alignment, which makes the crank nose longer.

Performance aftermarket crankshaft manufacturers offer many stroke options (this one is by Molnar and could be up to 4.050 inches). They also offer various lightweight options at extra cost with the same stroke. These are standard weight examples.

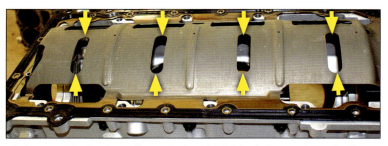

The bottom of the production windage tray has four slots: one over each bay. In this alignment, the four slots need to be lengthened about 1/4 inch toward the bottom pan rail. This can be done with a hand grinder with the tray off the engine.

The section bay (or bay number-1) of the stock windage tray needs to be straightened for use with a 4.00-inch crank. This straightening can be done with a small hammer.

Stroker Cranks

Scat, Callies, Manley, Eagle, and Molnar offer stroker cranks for the Gen III Hemi. The most popular tends to be the longer versions like the 4.00-inch crank used in the 426 Gen III crate engine. The increased displacement package is based around the 4.15-inch stroke and is the longest stroke to date. However, the 4.25-inch stroke unit is being developed right behind it. There are also 4.013-inch strokes for special racing applications and a 4.050-inch is also common. Caution: The 4.15 and 4.25 long-stroke cranks may require extra clearance be added to the block itself and require special rods matched to the crank.

4.000-inch stroker: The 3.72-inch "stroker" crank that's stock on the 6.4 Apache engine is easily installed into 5.7 and 6.1 engines. This stroker setup requires stock rods with special pistons to adjust for the .070-inch compression height.

The first true stroker crank, and one of the more popular, is the 4.00-inch stroker used in the aluminum-block 426 crate engine. It uses a standard 2.125-inch rod journal and a 6.00-inch length with a domed, high-compression-ratio Mahle piston.

These Gen III engines use a windage tray, but the stock tray needs two small modifications to clear the longer 4.00-inch stroke. The four slots in the tray over each journal must be elongated slightly, maybe 1/4 inch. The second mod is at the front near the number-1 rod journal on the side of the tray. The tray looks bent in at this spot and needs to be tapped out to match the other three bays.

4.050-inch stroker: Manley, Eagle, Scat, and others make a 4.05-inch stroker crank. It is similar to the 4.00-inch crank and requires the same windage tray modifications. The extra .050-inch of stroke gives you an additional 5 ci.

The Eagle and Manley rotating packages use a 2.00-inch rod journal and a rod length of 6.125 inches. The Indy Heads engine package with a 4.050-inch crank uses a 2.10-inch rod journal. Likewise, the piston's compression height can be used to adjust for the increased stroke rather than the rod length, or both can be adjusted.

For example, with the 4.050-inch crank and the 6.125-inch rod in the stock block (9.25-inch height), the piston's compression height is about 1.10 inches. You could switch to a 6.00-inch rod and the piston's compression height could be 1.225 inches.

4.125-inch stroker: The 4.125- to 4.150-inch stroker cranks are currently the biggest options available. This stroke isn't available in rotating assemblies and general performance catalogs yet, but by the time you read this, it will most likely be offered.

For these long-stroke cranks, you may need an aftermarket Milodon, Stef's, or similar windage tray, but you might be able to beat on the 4.00-inch tray a little harder in the center and sides to gain the extra .060-inch clearance. This package will use the 2.00-inch rod journals and probably a 6.00-inch rod length.

4.25-inch stroker: This package is being built one at a time using billet

Hemi-III Displacements					Hemi-III Displacements			
Stroke (inches)	Bore†† (inches)	Displacement (ci)	Base Engine		Stroke (inches)	Bore†† (inches)	Displacement (ci)	Base Engine
3.40*	4.06	352	NHRA SS		4.00	4.11	425	Stroker 6.4/392
3.40*	4.07	354	NHRA SS		4.00	4.12	427	426 Aluminum
3.40*	4.08	356	NHRA SS		4.00	4.13	429	426 Aluminum
3.40*	4.09	357	NHRA SS		4.00	4.14	431	426 Aluminum
3.58	3.92	345	5.7		4.00	4.15	433	426 Aluminum
3.58	3.93	347	5.7		4.00	4.18	439	426 Aluminum
3.58	3.94	349	5.7		4.05	4.09	426	Common Stroker
3.58	3.95	351	5.7		4.05	4.11	430	Common Stroker
3.58	4.00	360	Sleeved		4.05	4.13	433	Common Stroker
3.58	4.03	365	Sleeved		4.05	4.15	438	Common Stroker
3.58	4.06	371	6.1		4.05	4.18	439††	Common Stroker
3.58	4.07	373	6.1		4.13††	4.09	434	Stroker 6.4/392
3.58	4.08	374	6.1		4.13††	4.11	438	Stroker 6.4/392
3.58	4.09	376	6.2 (Hellcat)		4.13††	4.12	440	Stroker 426
3.58	4.10	378	6.2 (Hellcat)		4.13††	4.14	445	Stroker 426
3.58	4.11	380	6.2 (Hellcat)		4.13††	4.16	449	Stroker 426
3.58	4.12	382	6.2 (Hellcat)		4.25**††	4.12	453	Stroker 426 Modified
3.72	4.09	391	6.4 or 392		4.25**††	4.14	458	Stroker 426 Modified
3.72	4.10	393	6.4 or 392		4.25**††	4.16	462	Stroker 426 Modified
3.72	4.11	395	6.4 or 392		4.25**††	4.18	467	Stroker 426 Modified
3.72	4.12	397	6.4 or 392††					
3.795	4.06	393	6.1 with MP Stroker Crank					
3.795	4.07	395	6.1 with MP Stroker Crank					
3.795	4.08	396	6.1 with MP Stroker Crank					
3.795	4.09	398	6.4 with MP Stroker Crank					
4.00	3.92	386	Stroker 5.7					
4.00	3.94	390	Stroker 5.7					
4.00	4.06	414	Stroker 6.1					
4.00	4.08	418	Stroker 6.1					
4.00	4.09	420	Stroker 6.4/392					
4.00	4.10	422	Stroker 6.4/392					

Only .020 oversize, .030 not recommended

* The 3.40-inch stroke crank is used in the NHRA version of the supercharged 6.2 Hemi with the 4.09-inch bore for a race displacement of 354 inches.

** The 4.25-inch stroker crank requires the aluminum block be modified for clearance.

† The 4.05-inch stroke and the aluminum block's 4.185-inch bore (biggest sleeve) make a 440-ci displacement; the hardware is currently available.

†† The 4.125/4.13- and 4.25-inch-stroke cranks are not yet readily available but will be soon. Billet crankshafts are the only ones currently available in these strokes.

cranks but it will be available in rotating assemblies in the near future. I expect that this package will use the 1.89-inch rod journal and a rod length around 5.93 inches. This stroke will need an aftermarket windage tray.

Balance Holes

Typically, the first and last throw on the crank has a balance hole in it. It might be 1/2 to 3/4 inch in diameter. This may or may not be found on forged aftermarket cranks. This hole in the crank throw is part of the Gen III Hemi balance package also. These holes are not part of an external balance package.

All 3.58-inch-stroke cranks in Mopar small-blocks, including the A-engine 360s and the Magnum

On this Molnar crank, the various rod journals have large holes drilled through them, which are used as part of the crank balance package. These holes are typically 3/4 to 7/8 inch in diameter.

engine 5.9, used external balance weights on the dampener and fly-wheel/flexplate. Gen III Hemis do not use external balance.

Material

OEM production forged cranks are made from a heavy-duty alloy of mild steel that's usually 1050 and sometimes 1053. It is strong and easy to machine in very large quantities. This special heavy-duty alloy steel was designed for use in the forging process and offered a 65-percent increase in strength over 1010 or 1020 mild steel commonly used. Today, the aftermarket uses 4340 material that offers around a 75-percent gain over the 1050 base-line. (These strength numbers are based on SAE tensile strength data.)

Crank Prep

New or used, cranks need to be inspected, measured, and balanced, all part of basic crank prep. A new crank should come from the manufacturer ready to install, which means that your crank preparation simply consists of the machine shop measuring the various specifications.

The rod journals in the crank get oil from the mains; the welding rod at the upper right goes down to the main, bottom center. Each rod journal is oiled from a different main journal.

Balancing

Balancing is always required when building a max-performance engine. I strongly recommend balancing the crank, rods, pistons, and associated parts as an assembly. It is very important to balance the new rotating assembly. When re-building a production engine, I still recommend it although the machine shop may not require it.

Endplay

You can't really measure the crank's endplay until you install the crank into the block with bearings. Endplay is how much the crank moves forward and rearward in the main bearing caps. It is measured with a dial indication set up parallel to the crank's centerline and perpendicular to one of the counterweights.

Polishing

A new crank comes polished. In some cases, a used crank can be polished to remove very light scratches and normal wear before re-installation.

Repair

If the wear is high and/or there are scratches that are too deep to polish out, the crank is typically sent out for repair, which means that the shop has to grind it undersize.

This might be .020 inch. It is common to grind the crank .020/

The typical production crank uses undercut radii on each journal next to the flange or counterweight. Notice the relief or ring at both edges of the polished rod journal (center) and main journal (right). An aftermarket or performance crank uses a full-radius design, which will not be undercut. This change requires different bearings (radiused or narrow).

.020-inch undersize (mains/rods) and then use matching .020-inch undersize bearings. If the crank needs more serious repair and the amount of undersize grinding would be .040 or .060 inch, you should consider having the crank re-heat-treated by a process called nitrating.

Grinding the crank more than .020 or .030 inch, removes much of the crank's surface hardness. This hardness can be added back into the crank by the nitrating process. Smaller journals have weaker cranks, so I suggest limiting any grinding to .020 inch for max-performance applications and only use more heavily reground cranks in basic service.

Undersize Mains: Whether to grind a used crank undersize to repair the journals is a decision that you and the machine shop can make after the machine shop has inspected the crank very closely. Any grinding requirement is typically .010- or .020-inch undersize. This small amount of change does not affect its functional strength for performance applications. More than this and it could become an issue. You have more flexibility with a forged crank.

Crank Dampeners

The Gen III Hemi engine's basic vibration dampener design is a steel/cast-iron outer ring mounted to the

ATI currently offers the largest selection of SFI, racing, and performance vibration dampeners for the Gen III Hemi engine family. These units have different weights (13 to 6.45 pounds) and different outer diameters (8.9 to 5.67 inches). Note that smaller/larger diameters may require longer/shorter belts. Engines include stock (5.7, 6.4, 426), VVT, and non-VVT in cars, trucks, and the Hellcat (supercharged). ATI also offers overdrive and underdrive options. The 10-percent overdrive Hellcat unit adds 3 psi of boost pressure but requires extra clearance to the A/C bracket and a new belt. The underdrive unit's dampeners range in the 13- to 15-percent area; overdrive units are 6 to 18 percent.

All Gen III main bearings have two oiling holes in the upper shell. They are connected by a groove that does not extend to the ends of the shell.

The number-3 thrust bearing in Gen III engines is not the front and rear flange on the number-3 shell; they are these two flat, half-round inserts: one for the front of the number-3 main and one for the rear. The bulkhead and main and the cap are machined specially to accept these half-round inserts, which are held in place by the crank and the cap once the crank is installed.

hub by a rubber isolator (a thin piece around 360 degrees of the hub). The unique aspect of the Gen III dampener is that the outside diameter of the outer ring also serves as the drive for the fan belt or serpentine belt. This means that there are several grooves around the outside that guide and drive the front accessory drive (multi-groove belt). Gen III Hemis use the dampener as a front pulley in the serpentine belt front-drive system.

ATI offers SFI dampeners for the Gen III family of engines. ATI also offers over- and underdrive pulley/dampeners.

Front Cover

The Gen III Hemi family uses a cast-aluminum front cover, which covers the camshaft drive and seals the front of the engine (see Chapter 2 for more details).

Bearings

The main and rod bearings are offered in several sizes: standard, .010 inch, and .020 inch for undersize journals. Keep in mind that the number-3 bearing shell (the thrust bearing) has no thrust flange on either side. In place of the flange, an insert on each side of the main takes the thrust.

The inserts look like small, curved hockey sticks that are flat and half-round. The crank and the cap hold them in place. They must be installed before the cap is installed.

In addition, the standard bearings are fine for production, under-cut cranks. However, forged, high-performance/stroker cranks often use a full-radius journal and this feature requires radiusing the shells. These special bearing shells are called "radiused" or "narrow" depending on the manufacturer.

Crankshaft Specifications					
Engine	Crank	Stroke	Flange	Performance Stroke (inches)*	Material
5.7	Stock	3.58	Eight-bolt	3.58 to 4.05**	Forged
6.1	Stock	3.58	Eight-bolt	3.58 to 4.05**	Forged
6.4/392	Stock	3.72	Eight-bolt	3.72 to 4.05**	Forged
426	Stock	4.00	Eight-bolt	4.00 to 4.05**	Forged
All	Race	All	Eight-bolt	4.125 to 4.25	Billet***

* MDS and VVT are disconnected for these applications.
** The forged 4.050-inch cranks are readily available from Eagle, Molnar, and Compstar.
*** Currently the 4.125/4.13- and 4.25-inch stroke cranks are billet designs. This process makes them more expensive and that limits them to racing applications. This will be changed in the near future. A billet crank can be used in any performance application.

Any one of these packages could use a lightweight forged crank. However, neither a lightweight forged crank nor a billet crank is required. Cranks are mainly changed for displacement increase.

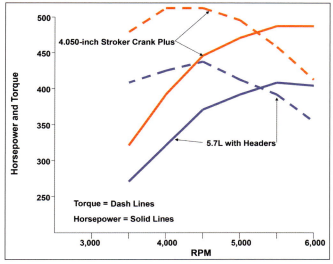

To illustrate the upgrade capability of the stroker kit, I selected a basic 410-hp 5.7 engine with headers because it is much easier to install headers than do a complete engine rebuild. The blue curves represent a stock 5.7-liter engine with headers. The red curves indicate the engine after a stroker crank (4.05 inches from Eagle) kit was installed. The cam was not changed. Assuming that the engine had the heads removed, they were upgraded to slightly larger intake valves. The resulting larger displacement (395 inches) increased the net compression ratio to 10.7 (from 10.2 in production). The resulting power increase was to 491, just about the same as the 6.4 (485 production), which uses a much bigger cam. A ported or larger throttle body was not used.

Standard bearing shells can be used, but they must be radiused by hand.

Crank Upgrades

The best crank tends to be based on the stroke length desired for the application. Today, forged crank manufacturers do not offer the extra-long strokes of 4.125 and 4.25 inches, so they are made in billet. This may change in the near future.

Aftermarket Options

Today, aftermarket manufacturers, including Callies, Eagle, Molnar, Manley, and Scat, make forged cranks in almost any stroke. The 3.40-inch crank for building the 354-inch NHRA Stock and Super Stock engines is made by the aftermarket. Likewise, the aftermarket makes the 3.50-inch crank used in the Canadian circle-track Pinty's series. In addition, Mopar Performance offers the 3.795-inch stroke

crank. Typically, an aftermarket crank would offer a longer stroke than the 3.72-inch production crank, which is why they are called strokers but shorter cranks can also be made.

Aftermarket cranks, such as those from Callies (Compstar), Eagle, Scat, Manley, and Molnar, all have similar features and benefits. They all go up to 4.050 inches and may or may not go down to the stock stroke of 3.58 inches. The first one to offer a 4.125-inch or so crank will have the advantage in the marketplace.

Cost is the one comparative factor that takes a little more research. Complete, or rotating assembly, packages offered by these companies usually include crank, rods, pistons, main and rod bearings, and piston rings. This type of package can offer savings over buying each individual part from different suppliers. As these packages become better defined, they will probably be offered by the high-volume catalog centers, such as Summit Racing and Jegs.

Billet Cranks

Billet cranks are popular in racing. They offer the ability to use a crank that has a unique stroke, and one that is not readily available as a forged crank. Today, the 4.125- and 4.25-inch cranks for the Gen III family of engines are basically billet cranks. Moldex, Scat, and Winberg are aftermarket companies that offer billet crankshafts. Most manufacturers can grind these cranks on a special-order basis.

Typically a forging has limits in stroke, such as 3.58 to 4.050 inches. There are no limits on the billet version, so they are best for custom applications, custom material, unique spacings, etc. The drawback is that they tend to be much more expensive than forged designs.

Main Bearings

Never grind the crank undersize until you know that bearings are available for the size you have selected. Standard bearings are readily available and .010- and .020-inch bearings for cranks ground to these undersizes are also readily available. It is rare to use a size smaller than .020-inch under.

Rear Seal

The rear flange on the Gen III Hemi cranks is wider than standard and machined perfectly smooth. The rear seal is a 360-degree seal and seals against the crank's rear flange. This seal fits over the end of the crank and is held in place by an aluminum retainer that bolts to the block. Almost all previous V-8 (small- and big-blocks) rear seals were two-piece.

Connecting Rods

The engine's connecting rods connect the piston to the crank and are therefore very important. Because

The production I-beam rod has a screw for cap attachment, rather than a bolt and nut design. This rod has a pin bushing (right), but some production rods use no bushing but rather a pressed pin in the small end.

This is the standard Gen III connecting rod cap; note the straight, flat bridge from the upper rod bolt to the lower rod bolt once it leaves the bolt area.

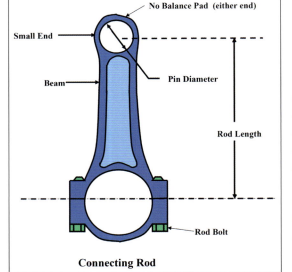

Connecting Rod

Gen III Hemi rods are unique among production rods because they do not have balance pads on either end, and therefore they are similar to aftermarket rods. These rods use screw retention rather than bolt and nut retention, which is also somewhat rare in production.

This heavy-duty production rod is supposed to come with VVT engines but may not. Note the tabs pointing forward and rearward in the middle of the bolt bridge.

All production rods are cracked-cap designs, which means that each rod cap is cracked or broken off from the beam, and that makes the cap's parting line unique. This feature means that the caps cannot be interchanged.

Production rods use both pressed pins and floating pins. It was supposed to be fairly straightforward: the 5.7 engine used pressed pins and the 6.1 and 6.4 engines used floating pins. However, pressed pins are found in 6.4 engines and floating pins are found

5.7 Hemi Pressed Pin

6.4 Hemi Floating Pin

in some 5.7s, so double-check to verify which style your engine has. If the piston has lock rings on the ends of the pin, it has floating pins (pressed pins do not use any pin locks). Aftermarket pistons are built to include piston locks. The owner can choose to omit them for a pressed-pin application.

the rod rotates with the crank, it is highly stressed during normal operation. Before purchasing, several rod aspects need to be considered: material, style, cap, and pin retention along with overall weight.

Production Rod

Production Hemi III connecting rods are made of powdered metal and feature a cracked-cap design. The advantage to the cracked cap is that the connecting rod cap can only be installed one way and only on the rod that it came from. To date, production has built two rods: standard and VVT. The VVT rod's cap has small "tabs" that stick out left and right of the strap-ridge stiffener between the two-rod bolts. As of this writing, not all VVT engines use the upgraded rod. The VVT rod is stronger. Gen III Hemi rods are .933-inch wide.

Do not number stamp or punch the connecting rods or caps for identification because this may damage the rods. Instead, you should mark the rods and caps with a permanent marker or scribe tool.

Rod Length

The rod length is defined as the distance from the center of the big end of the rod (crank journal) to the center of the small end of the rod (pin end). The production Hemi III engine uses two lengths that include 6.242 inches for the 5.7 and 6.1 and 6.200 inches for the 6.4. The 426 crate engines use a 6.00-inch length and the Canadian circle-track package (for the Pinty's series) uses a 6.30-inch length.

Stroker Rod Length

The 392 has a 3.72-inch stroke and the 5.7 and 6.1 have a 3.58-inch stroke, so if you install a 392 crank in the 5.7, the piston would stick out of the block by .070 inch. Note that the 392 rod is .042-inch shorter. The other .028 inch is adjusted for in the piston's compression height.

In some cases, moving the piston up in the cylinder bore is desired to increase the engine's compression ratio. On longer-strokes such as the 426 and its 4.00-inch stroke, the rod

The obvious approach is the rod bolt and nut. However, in performance rods and racing rods, it is much more popular to use a bolt that threads directly into the beam of the rod with no nuts. Typically, when this is done, the hole is drilled through. The attaching bolts are also 12-point (aftermarket). The Hemi III rods use the "race" rod cap retention method of the bolt going directly into the beam; no rod nuts. The bolts used in production are 6-point.

Connecting Rod Specifications					
Displacement	**5.7**	**6.1**	**6.4**	**6.2**	**426**
Center-to-Center Length (inches)	6.242	6.242	6.200	6.200	6.000
Weight (grams)	597	597	658	658	N/A
Bolt (points)	12	12	12	12	12
Rod Width (inch)	.933	.933	.933	.933	.933
Bearing Bore ID (inches)	2.125	2.125	2.125	2.125	2.125
Pin ID, floating (inch)	.9456	.9434	.9434	.9434	.9434

Rod Length Measurement

The rod length, or center-to-center distance, is easy to see and you can estimate it with a ruler. All 5.7 Gen III engines use a rod length of 6.240 inches (6.200 on the 6.4 and 6.00 on the 426). Although it is easy to see, it is very difficult to measure accurately. The difference between 6.00 and 6.20 inches, for example, is very important. You can use this formula to calculate rod length:

$$RL = PD \div 2 + BD \div 2 + BL$$

Where:
RL = rod length, center-to-center distance
PD = pin diameter
BD = big-end diameter
BL = beam length

For example, let's assume that our measurements are as follows: pin bore diameter of .946-inch, big-end bore diameter of 2.125 and a beam length 4.7045 inches.
RL = .946 ÷ 2 + 2.125 ÷ 2 + 4.7045 = 6.240 inches ■

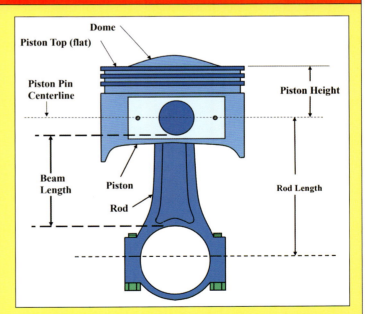

Rod length is very important, along with the piston height, in stroker calculation and resulting compression ratio calculations.

This Compstar H-beam high-performance rod has a bushing for a floating pin and a pin oiling hole at the top of the side-slot, or H. There is one of these pin oiling holes on each side of the rod.

is shorter at 6.00 inches to help keep the top of the piston below the deck surface. With a longer-stroke package, you must look at the rod and the piston as a team with one of their goals to keep the top of the piston below the deck surface.

Beam Types

The I-beam and the H-beam are the two connecting rods styles. Production rods are I-beam style while many aftermarket rods are H-beam style. Heavy-duty I-beam rods are offered for high-end racing application.

Rod Weight

The connecting rod's weight is very important to the engine's perfor-

mance. Chrysler engineers optimized beam stiffness while minimizing rod mass, or weight. For example, the 5.9 small-block engine used 6.123-inch rods that weighed about 758 grams; the heavy-duty 6.4 Gen III Hemi rod weighs 658 grams, or 100 grams lighter, in the same basic length. A 6.4 Gen III heavy-duty rod weighs about 469 grams on the big end and 189 grams on the small end.

Inertia

The inertia of a rotating part is directly related to its weight, and lightweight is better. For example, it is more important to remove weight from the big end (the 469 grams) than from the small end (189 grams) because the small end doesn't rotate around the crank.

Rod Material

Today, if you want a stronger rod, the material that it is made from is very important. A typical aftermarket forged rod is stronger than a production rod, even if the rod bolts are replaced with high-strength steel bolts.

Rod Ratio

An engine's rod ratio is the rod's length (center-to-center distance)

divided by the crank's stroke. All 5.7 Gen III engines share the same 1.74 rod ratio; earlier 360/5.9 small-blocks were about 1.71. The 6.4/392 Gen III engines share a 1.67 rod ratio; the 426 Gen III or 4.00-inch crank has a 1.56 rod ratio. The 4.00-inch crank in small-blocks has a rod ratio of 1.53.

It can be expensive to change both the rods and the pistons to obtain a new rod ratio.

Rod Journal

All rod journals in production engines are 2.125 inches. This is true because there is no reason to change it. Even the long-stroke 4.00-inch crank fits with no major clearance issues.

However, as the stroke gets longer, there is a clearance issue with the sides of the block. Yes, you could add notches to each side, but that is an expensive machining process. To make everything a bolt-in, aftermarket crank manufacturers make a small change: They drop the rod journal size to 2.00 inches and make the crank journal to match. This allows the outside edge of the rod on the crank journal to be closer to the center of the journal and therefore creates more clearance to the sides of the block. Otherwise, you would probably have

Rod upgrades come in all sizes and shapes as shown by this Scat I-beam rod. Typical of the aftermarket rods, they can be made with or without pin bushings so they can be used with either floating-pin or pressed-pin rods.

Manley offers many styles of connecting rods for many applications. This is Manley's standard I-beam rod for Gen III Hemis. The H-beam Gen III Hemi rod appears in the floating-pin package. (Photo Courtesy Manley Performance)

This K1 H-beam rod features the typical performance pin bushing and cap-bolt retention desired in a performance rod. Manufacturers can make these rods in several lengths, including 6.00, 6.200, and 6.240 inches. (Photo Courtesy K1)

Eagle offers this H-beam forged-steel rod for Gen III Hemis; it uses 12-point rod screws to hold the cap on. Several length options are offered: 6.00, 6.20, and 6.24 inches.

Aluminum rods look so much thicker and beefier than steel rods, but they are actually lighter. This aluminum rod from MGP weighs less than a popular steel rod. Aluminum rods are typically not used in street engines but are often recommended on very high-output supercharged engines.

to buy new rods with the crank, so this is a great adjustment and doesn't add any expense in most cases.

Companies that sell rotating assemblies, including Manley and Eagle, use the 2.00-inch journal with their 4.050-inch cranks. Indy Heads uses a 2.100-inch journal with its 4.050-inch crank packages. In addition, it appears that the 2.00-inch journal will be standard with 4.125 and 4.15 cranks. The even-longer 4.25-inch stroke package is based on the small 1.89-inch rod journal.

Rod Upgrades

The first upgrade is to install upgraded rod bolts. The next upgrade is to use an H-beam rod, which is offered by almost all manufacturers.

Rod Prep

With powdered-metal production rods, there isn't much that you should do to prep them and if you source new rods from the aftermarket, they should come ready to be installed. You and the machine shop may check them over but probably not do anything to them. The exception might be replacing the rod bolts in the stock rods with ARP high-strength bolts.

Typically, rod side clearance isn't measured until the rods are assembled onto the crankshaft. Then the clearance is measured between the two adjacent rods on the same journal. This clearance can be estimated by measuring the thickness of the rods and the width of the rod journal on the crank.

Chances are that if you are going to rebuild the engine using a stroker crank, you will also add other performance parts as well. The stock 6.4/392 Hemi engine makes 485 hp. Therefore, if you install a 4.00-inch stroker crank (from Molnar) into your package, you could also add headers (from American Racing Headers), a bigger hydraulic cam (from Comp Cams), CNC-ported heads (same valve size; from Modern Muscle, Arrow Racing, or Modern Cylinder Head), and a 4-barrel throttle body (from Edelbrock) with a high-rise intake manifold (from Arrow Racing). Together, these changes would make about 649 hp, or a gain of about 160 hp over stock. Although you are supposed to change one thing at a time, it is not realistic to install the 4-inch crank without adding other performance items at the same time. It may not be a good test but I think it is representative of what you might do and the gain you would get.

Graph: Horsepower and Torque vs. RPM

- 6.4 with 4.00 Crank, CNC-Ported Head, 4-Barrel Throttle Body, Big Cam, and Headers
- 6.4 Stock (485-hp Peak)

	Connecting Rod Upgrades			
Engine	**Rod**	**1st Upgrade: Rebuild**	**2nd Upgrade: Performance**	**Race**
5.7 and 6.1	Stock	Upgraded Bolts	High-Performance H-Beam	High-Performance I-Beam or H-Beam
6.4	Stock	Upgraded Bolts	High-Performance H-Beam	High-Performance I-Beam or H-Beam
426*	High-Performance H-Beam	No Need	No Need	No Need
* The 426 comes with high-performance H-beam rods including good bolts so upgrades are not required.				
Aluminum (from BME and MGP) and/or titanium (from Manley) rods can be added to any package but are not required.				

PISTONS AND RINGS

The piston sits in the block's cylinder bore, the rod is attached via the pin, and the piston top seals the combustion chamber. This means that there are design aspects of the piston that affect other parts of the engines. The piston pin is designed to hold the piston onto the connecting rod while the piston moves up and down in the bore and the rod swings around the crank.

Typically, the max-performance piston choice is between hypereutectic and forged. Because Gen III Hemi

pistons are hypereutectic, the choice becomes easier. Production blocks are basically thin-wall castings so the amount of overbore allowed is very limited: .010 or .020 inch. On the other hand, a fairly large selection of production bore sizes is available from 3.91 inches on the 5.7 to 4.09 inches on the 6.4. High-performance aluminum blocks offer even more options by changing sleeves.

Pistons

The piston's primary job is to seal the cylinder's combustion chamber while the crank rotates. The piston rings are the key piece that allows the piston to do this job. The rings must seal to the piston and they must seal to the cylinder wall as the assembly

The top of the stock piston has a notch to the left (exhaust) and one to the right (intake). The rest of the piston is flat, but it has a slight dome with a dish toward the center. The valve notches in the stock piston indicate how close the piston is to the head in piston-to-head clearance and valve-to-piston clearance. Remember, even the 5.7 has a 10.2:1 compression ratio rather than the older engines' 9:1 and 9.5:1. That also requires a lot of clearance.

The Wiseco piston has a closed combustion chamber shape for the crown. The skirts are coated with a friction-reducing material. This piston has large valve notches, so it probably has clearance for large camshafts. (Photo Courtesy Wiseco)

This production piston looks more like a race piston than the typical production piston. The overall piston height is very short. The rings are very thin (1.2 mm) and the ring package is very compact (short). The piston skirts are coated with an anti-friction coating.

moves up and down. The piston must be strong enough to support the high pressures in the cylinder's combustion process and yet light enough to allow the engine to perform at its best.

Production Pistons

Gen III Hemi production pistons are lightweight designs made from a hypereutectic aluminum alloy. Most production compression ratios (CRs) are in the range of 10.2 to 10.7, except for the 6.2 Hellcat engine (supercharged) that has a 9.5 CR. You should always plan to replace the pistons and rings with every rebuild.

The cylinder bore in the block wears with average use and mileage. To get the bores straight and round again and remove the damage done by the wearing process, the machine shop must bore and hone the cylinder bores to larger numbers. If the original pistons are not replaced, the clearance between the piston skirt and the cylinder wall increases. This increased clearance increases the engine's blow-by and hurts sealing, which reduces horsepower. A large-diameter piston solves these problems.

Production pistons come in pressed-pin and floating-pin styles. Supposedly the 5.7 and 6.1 engines received the pressed pin pistons and the larger 6.4 engines received the floating pin pistons. That isn't the case because there are 6.4 engines with pressed pins and 5.7 engines with floating pins. The aftermarket solved this situation by building the piston with

lock grooves on each end of the pin and including the pin and locks with the piston. If you have a pressed-pin engine, the locks are left off.

Mahle makes most of the production pistons, and it also produces pistons for the aftermarket. Although somewhat small, these pistons also have a dome on top. In some cases, the manufacturer adds two small, rounded cups on each side of the combustion chamber for the dual spark plugs to help propagate the flame travel in the combustion chamber.

Piston Cooling Jets

A unique feature of Gen III Hemis is that the piston cooling jets are located at the top of the crankcase between the banks; they squirt oil directly onto the underside of the pistons. The cooling jet has two arms and a center post. One arm squirts oil on each bank. The post presses directly into the main oil galley above the crank and below the cam; a small bolt holds them. If they are not used, the hole must be plugged. Although most Gen III Hemis seem to have these cooling jets, they are omitted from some models.

These pistons' cooling jets must be removed if long-stroke cranks are used: 4.00 inches and up. They must be replaced with a plug for the holes.

Piston Weight

Chrysler engineers worked hard to reduce piston weight. They know all the good things that less weight

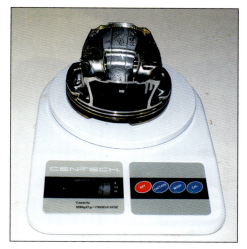

The stock 6.4 piston (4.09-inch bore) weighs about 437 grams. The 5.7 piston is even lighter at 413 grams (smaller bore). That is very light for a production piston; it's more like a race piston (NHRA Super Stock).

and less inertia do for overall engine performance. The Gen II 426 Hemi used a piston that weighed about 850 grams, but it had a 4.25-inch bore. The Mopar small-block 5.9 piston (Magnum family from 1993 to 2003) was the best design of the early small-blocks and it weighed 470 grams at 4.00 inches. The earlier 360 version used a 584-gram piston. The 6.4 Hemi piston weighs 437 grams at a bore size of 4.09 and the 6.4 piston has a slight dome. The 5.7 piston weighs only 413 grams but is based on the smaller 3.91 bore size.

One advantage of forged aftermarket pistons is that they can be lighter than production pistons. For example, Diamond Pistons has some 5.7 pistons that weigh about 390 grams. Note that manufacturers often indicate "call" in place of the actual piston weight because they can offer custom machining, which changes these numbers.

Stroker-crank pistons tend to be shorter than stock pistons and this

Piston Specifications						
	5.7	6.1	6.4	6.2	426	354
Nominal bore size (inches)	3.917	4.055	4.09	4.09	4.125	4.09
Piston weight (grams)	413	435	437	502	N/A	437
Piston height (inches)	1.21	1.21	1.18	1.18	N/A	N/A
Pin bore diameter (inch)	.946	.946	.946	.946	.946	.946

helps reduce piston weight. Because of the high loads encountered, the Hellcat, or supercharged piston, is slightly heavier and therefore much stronger, which is highly recommended for any supercharged engine.

Inertia

Weight relates directly to inertia. A lighter piston weight means less inertia and that means less force on the rod and crankshaft. Less inertia is generally good for performance, especially acceleration or changing speeds.

Height

The piston height is the distance from the center of the pin to the top (flat) of the piston. The 5.7 Gen III Hemi is 1.21 inches (the earlier 5.9 small-block was 1.57 inches). Both the 5.7 Hemi and the 5.9 wedge small-block use a 3.58-inch-stroke crank. This indicates how much shorter the new Gen III engines are than their small-block predecessors.

A ring package includes a top ring, a second ring, and an oil ring, which has three pieces (two rails and an expander). All fit into the .763-inch height. The first two rings are 1.2 mm while the oil ring is a 2-mm (early engines were 1.5 mm). That is much thinner than the 1/16-inch rings that were used in almost all performance applications a few years ago.

Gas Ports

Gas ports are small holes drilled into the top of the piston to allow cylinder gas pressure to push the ring against the cylinder wall for better sealing and higher speeds. Gas ports are generally used with thin, light rings such as .043 or 0.47 inch.

The gas ports are 14 equally spaced small holes drilled vertically from the piston top into the ring groove. These small holes are .040 to .045 inch in diameter. These holes should be drilled by the machine shop or by the piston manufacturer. Gas ports work, but you should be aware that high cylinder loading causes rapid cylinder bore wear. Gas ports are used in racing but I do not recommend them for street use.

Ross Pistons offers this domed Hemi piston for use with gas ports. You can see the small, vertical holes drilled around the outside of the flat. They are drilled down to the backside of the top ring land. I do not recommend them for street use, but they are popular in drag racing.

Piston Skirt

The piston skirt is the bottom part of the piston below the pin. It holds the piston upright in the cylinder bore and guides the piston up and down the cylinder walls.

The skirt transfers loads to the cylinder walls as the piston moves up and down. Gen III pistons have a black or dark-colored coating on the skirts, Molykote, which is considered a solid lubricant. You might also call it a friction reducer.

Ring Grooves

Typically, three ring grooves are cut into the piston: 5/64 inch, 1/16 inch, 1.5 mm, 1.2 mm, or .043 inch. If you want to use a specific ring thickness, you must tell the piston manufacturer, so it can cut the correct grooves. Otherwise, when you receive the pistons, you must match the rings to the grooves. In some cases, this could limit your selection. Gen III engines from 2003 to 2008 used 1.50-, 1.50-, and 3.00-mm rings. The 2009 and newer Gen III engines use a 1.20-, 1.20-, 2.0-mm ring set.

Piston Top

The 5.7 piston top looks flat but has a very slight dome. Gen III Hemis do not use dished pistons to date except for the Hellcat and the

This Mahle piston is basically flat except for the two somewhat large valve notches on each side of the piston. With performance pistons, several piston bore diameters and several piston heights can be made so they can service all three engines: 5.7, 6.1, and 6.4.

long-stroke versions. The exceptions may be the newer Ram truck and Jeep engines as more SRT and Hellcat versions evolve into these areas.

Compression Ratio

Gen III Hemi pistons have production compression ratios in the 10.2 to 10.7:1 area, which is higher than the compression ratios built into the small-block pistons of the earlier era (1992–2003), which were around 9.2 to 9.5:1. This is related to computer speed and sensor accuracy along with the two knock sensors. With the Hemi combustion chamber design, the intake valve is on one side of the chamber and the exhaust valve is on the other side of the chamber, which makes the chamber wide. Gen III Hemis use a shallow chamber in the head. The result of all this is that the piston is close to the head. This piston-to-head clearance should be measured closely.

The exception to the 10.2 to 10.7:1 compression ratio is the 2014–2017

Hellcat, which is built at 9.5:1. When supercharging an engine, it is advantageous to lower the compression ratio from the naturally aspirated versions.

CC-ing Procedure

The process of cc-ing is to measure the volume of a given space. It is commonly used in measuring the volume of the cylinder head's combustion chamber. However, it is also used to measure the volume of a domed piston and/or pistons with valve notches. Because most Gen III Hemi pistons have a small dome and valve notches in some cases, this procedure is very important. It enables you to calculate the compression ratio accurately.

To perform the procedure, you bring the piston to TDC measuring with a dial indicator on the top of the piston (the indicator is zeroed). Drop the piston down the bore exactly .300 inch below TDC; seal the rings to the cylinder wall with a

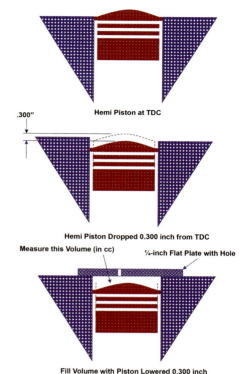

To cc the effects of a piston dome, a dish, or valve notches, you must measure the volume. To measure the volume of a shallow chamber design, you need to lower the piston .300 inch below TDC to be sure that the top is below the deck surface.

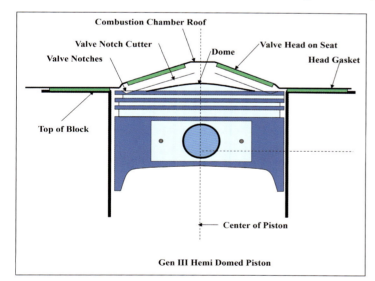

Even production pistons are close to the cylinder head or combustion chamber. The 5.7 cam is larger than many older production cams (non-performance), but it is not large. It is very important in these Gen III engines to measure piston-to-head clearances and piston-to-valve clearances. The chambers are very shallow for the hemispherical chambers.

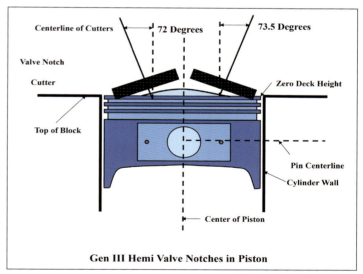

Shallow-chamber Gen III Hemi heads use different valve angles than other Hemi engines. They are much closer to wedge valve angles but certainly not in usual wedge locations. There is no "F" on Gen III Hemi pistons: The intake valve notch must always go on the top and the exhaust valve notch goes toward the outside of the block.

Measuring Piston Head Volume

Whether you have a dish volume, a dome volume, or valve notches, there is no easy way to find the exact volume without cc-ing. The procedure is basically the same as cc-ing a cylinder head's combustion chamber.

Most Gen III Hemi piston domes have a height of .15 to .25 inch, so the .300-inch down distance was selected based on the typical height of the dome being lower. The top of the dome must be below the deck surface.

Here's an example: Assume that you have a 6.4 engine with a 4.09-inch bore. The perfect .300-inch down volume is 64.59. Therefore, if you have a piston with a 5.0-cc dome, the measured volume should be 59.59.

If you have a piston with a 10-cc dome, the measured volume should be 54.59.

If you have a flat piston with intake and six exhaust valve notches, the measured volume should be 70.59. ■

Engine	Exact Bore (inches)	.300-inch Down Volume (cc)
5.7	3.917	59.24
	3.927	59.55
	3.937	59.85
	3.957	60.15
6.1	4.055	63.49
	4.065	63.80
	4.075	64.12
	4.085	64.43
6.4	4.090	64.59
	4.100	64.91
	4.110	65.22
	4.120	65.54
426 inches	4.125	65.70
	4.135	66.02
	4.145	66.34
	4.155	66.66
	4.165	66.98
	4.175	67.30
	4.200	68.11

The .300-inch down fill volume procedure is not a function of total displacement or stroke but only bore size.

light grease such as Vaseline. Wipe off any excess above the top of the piston. Run a small bead of light grease around the top of the cylinder bore and press your cc-ing plate (with a hole in it) against the deck surface over the bore.

You add the fluid to the burette (measures the amount of fluid used) and move the burette over the hole in the plate; the fluid is dispensed into the space above the piston. Keep the fill hole on the high side of the cylinder. Fill the volume above the piston with fluid.

Record the volume and compare it to the perfect .300-inch disc.

Compression Ratio

The block, crank, rod, cylinder head, and head gasket play a large

This Mahle high-compression piston has a big dome, large valve notches, and large reliefs for the spark plugs to help flame travel.

role in determining the engine's compression ratio. However, the piston's part is the most significant and the piston's dome design helps dictate the compression ratio.

The piston affects the volume above the piston at TDC by changing its compression height; it can also affect this volume by having a dome

or dish as part of the top's shape. This compression ratio discussion could be placed in Chapter 7 or Chapter 3, but I placed it here because the piston can change the compression ratio once the engine specs have been defined. If the engine has too little or too much ratio, the piston can fix it. This adjustment is not so easy with the other parts and using the piston is the most cost-effective approach.

The engine's compression ratio is the volume above the piston at BDC (VBDC) divided by the volume above the piston at TDC (VTDC). The VBDC is actually the VTDC plus the cylinder's displacement. The VTDC is the sum of the combustion chamber volume (CC) plus the head gasket volume (HG) plus the deck height volume (DH) plus the piston's dome/dish volume including notches (DV).

For example, let's assume that you have a 6.4 engine that's .020-inch oversize (4.11 inch) with a 4.00-inch crank (402 ci), a 70-cc chamber head, a .040-inch-thick gasket, a .030-inch below-deck piston height (or deck height), and a domed piston with a 5-cc dome with 8 cc in valve notches.

One cylinder displaces 53.068 ci, or 869.62 milliliters (53.068 x 16.387). The head gasket has a volume of 8.70 and the deck height has

a volume of 6.51. So the VTDC is 88.21 (70.0 + 8.70 +3.00 + 6.51).

To find compression ratio, use this formula:

$$CR = VBDC \div VTDC$$

Continuing the example, the CR is 10.85:1 ([869.62 + 88.21] ÷ 88.21 = 957.83 ÷ 88.21).

Supercharged Compression Ratio

A supercharged engine generally does not have the same compression ratio as a naturally aspirated engine. Race cars solve this problem by using expensive racing gas, which has a much higher octane level than typical pump gas. The 707-hp Hellcat supercharged engine uses a 9.5:1 CR rather than the standard engine's 10.7:1.

About 20 years ago, there was a big performance push for superchargers on the street (as today) because of the easy performance gains. They were added to production 9.5:1 engines and the head gaskets failed,

This Mahle piston for Gen III engines almost looks flat, but the open chamber relief is actually a 12-cc dish. The valve notch is at the top.

Not all Gen III pistons are flat or domed on top. The Hellcat uses a dished piston. This CP Pistons design mirrors the closed chamber used on these heads with a dished drop in the top of the piston. Even with the dish, you can see the valve notches (lower left).

Calculating Piston Speed

Piston speed is calculated by using this equation:

$$PS = 2 \times RPM \times S$$

Where:
PS = piston speed (fpm, or feet per minute)

RPM = engine speed
S = stroke (feet)

The mean piston speed for the 5.7 engine is about 3,900 fpm. The mean piston speed for the 426 Gen III engine is about 4,000. ∎

Calculating Piston Compression Height

One of the ways to change an engine's compression ratio is to move the piston up or down by changing its compression height (distance from the center of the piston pin to the top of the piston), which changes the piston's deck height.

If you do not know your piston compression height, you can calculate it by re-adjusting the block height equation from Chapter 2 and solving for compression height.

$$CH = BH - S \div 2 - RL - DH$$

Where:
CH = compression height
BH = block height
S = stroke
RL = rod length (center-to-center)
DH = deck height

Continuing the example for the 5.7 block height: The stroke is 3.58, block height is 9.25, piston height unknown, piston deck height is .010, and rod length is 6.24. That makes the compression height 1.21 inches (9.25 - 3.58 ÷ 2 - .010 - 6.24). ∎

or worse. The solution was to install a thick head gasket and reduce the ratio about one full point.

That does not seem to be a solution today. Why? It appears that the dual knock sensors used on Gen III Hemi engines is saving everyone's engine by discreetly pulling back the spark advance so that the engine doesn't hurt itself. I recommend that you drop the ratio one point if you are building an engine, but it seems the knock sensors will save the engine anyway. Caution: Do not modify or eliminate the knock sensors program in your computer with any reprogramming.

Piston Upgrades

Many piston manufacturers sell a variety of pistons for the Gen III Hemi.

The production piston is quite light and all manufacturers offer lightweight options for extra cost to cover the necessary added machining time. Lightweight pistons are a plus but not required. The production 1.2-mm rings are very thin.

Diamond makes this piston for the 5.7 that comes with domes from 6.3 to 14.2 cc. The open-chamber mirror for the slight dome is just barely visible. It is light at 412 grams (with 4.050-inch stroker) and has large valve notches. The dome height may just balance the volume of the notches. (Photo Courtesy Diamond Racing Pistons)

Piston Pin Options				
Engine	Pistons*	1st Upgrade**	Rings (mm)	2nd Upgrade***
5.7	Pressed Pin	Floating Pin	1.2 and 1.5	Oversize
6.1	Pressed Pin	Floating Pin	1.2 and 1.5	Oversize
6.4	Pressed Pin	Floating Pin	1.2 and 1.5	Oversize

* The 5.7 pressed-pin and 6.4 floating-pin rule is not holding true. If you have a pressed pin with either engine, upgrade to floating pins.
** Also requires a floating-pin rod.
*** Any engine rebuild requires oversize pistons; try lighter weights for the upgrade. Piston manufacturers generally do not publish the weights of high-performance pistons.

Most manufacturers offer a piston skirt coating similar to the Molykote used on production pistons. They all offer floating pin designs. The block (bore size), crank (stroke), and CR are the keys, not any specific feature.

If you're building a high-horsepower engine, you should

Aftermarket piston manufacturers make several versions of the Gen III Hemi piston. They come in the popular bore sizes and the readily available piston ring selections. Ordinarily increasing the engine's CR a point or two is a common upgrade. I have not done that with Gen III engines because they are already high in CR, at 10.2 to 10.7:1. If you push the ratio higher, that means higher-octane fuel, which is race gas, which is expensive. I try to stay within 1/2 point of the stock ratio. This Diamond piston for the 6.4 shows the dome for the closed chamber used on the 6.4.

upgrade to a floating-pin piston. Then you should determine if the manufacturer offers the piston, the required bore size, and the CR. They are typically in the same price range; remember, more money for lighter pistons.

The real deciding factor is availability. High-performance street and street/strip versions are basically the same price. Full race is different; you will pay more for lighter pistons.

Aftermarket Offerings

The aftermarket makes pistons in almost any size and shape. Production pistons are hypereutectic; you only have forged and hypereutectic designs to consider. Most manufacturers offer forged pistons in the

Diamond's forged piston for the 6.1 engine has two valve notches, one on each side of the slight dome. It is also offered with a 7- to 26.5-cc dish. It is as light as 407 grams with the 4.050-inch stroker. (Photo Courtesy Diamond Racing Pistons)

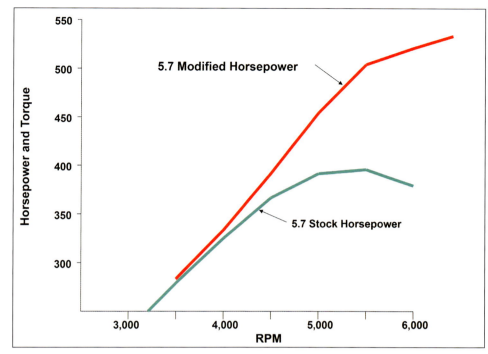

If you took a 5.7 engine apart to rebuild it, you would probably add a few things to the performance package as you put it back together. I selected headers, a ported Modern Muscle throttle body, a bigger Modern Muscle or Comp Cams cam at .550-inch lift, a big valve (2.05 inches), a CNC-ported head (Modern Muscle or Modern Cylinder Head), and a slight increase in CR to 11:1 by using domed Mahle or Diamond pistons. I compared this high-performance package to the stock 5.7-baseline engine at 385 hp. The modified engine made about 537 hp.

production bore sizes and common oversizes. Manufacturers, such as Mahle, Icon, Diamond, Wiseco, JE, Modern Muscle, CP-Carillo, TRW, and Ross, offer forged pistons in many configurations based on the stock engine's CR as the bottom line.

The pin heights and the ring package are the most popular things to change or adjust followed by dome height and shape. When selecting a piston, weight is a key factor if all other aspects are equal. As you would expect, a lighter piston is more expensive. When it comes time to choose a piston, you need to find one that fits the bore size and CR of your build.

In 2015–2016, only one manufacturer offered billet pistons, Gib-Tec.

Billet pistons are often selected for racing applications and not required for street engines. Billet pistons are not related to horsepower and I consider them a "pro" race part.

An application for racing billet pistons would be to make a 300-gram piston and see if it lived. Do not print that. When I wrote this I though it would be a race-only part, but I don't think that is true anymore. I have found that the Jeep Grand Cherokee SRTs and the FCA all-wheel-drive hot rod since 2010 or so have superchargers and make ridiculous amounts of horsepower, upwards of 1,000 on the street.

They break stuff, including pistons with 10.7 CR and they have too much CR for the boost, the Hell-

cat has a 9.5 CR. The Hellcat has a 4.09 bore while many of the 6.1 SRT engines have only 4.05. There are no 9.5 CR pistons. It's the perfect spot for a billet piston.

This means that you can design your own piston and someone will make it for you. To look at it another way, perhaps you see a feature on a Mahle piston that you like and another feature that you saw on another company's race piston. The billet piston offers you the ability to put the two features together for your engine. See something on the 5.7 that's not on the big-bore 426 Gen III crate engine? Billet machining is the solution.

Piston and Ring Prep

Like most V-8 engines, Gen III pistons can be purchased ready to install; in many cases, your machine shop makes the final valve-notch cuts. Most of the rest of the piston and ring prep comes down to measuring clearances and gaps because the manufacturer did the real prep.

Cylinder Bore

The Gen III finished bore size is one of the most important numbers relating to the short-block. Typically, the machine shop does not finish-hone the cylinder bores until they have the finished piston in hand. Then they measure the new piston's exact bore size and add the desired clearance; that is the number that they use for the finished hone size.

Piston-to-Wall Clearance

This clearance is usually accomplished by the machine shop as part of the boring and honing process. The piston manufacturer provides the piston clearance specification.

The piston-to-cylinder wall clearance is usually measured by the machine shop when they bore and hone the block. The piston bore size is measured across the skirt below the oil ring. Each piston manufacturer recommends the best clearance for that piston.

Deck Height

The deck height is measured from the top of the piston to the top of the block using a dial indicator. Once the Gen III block has been bored and honed, the piston and rod can be partially assembled (no rings required) and the piston's deck height can be checked. This measurement is very important to the engine's CR calculation. If you have too much CR or not enough, now is the time to find out because you can still do something about it. The 5.7 Hemi's piston height (center of pin to top of piston) is 1.21; the earlier 5.9 small-block was 1.57. Both use a 3.58-inch stroke, which illustrates how much shorter the Gen III block is.

Ring Gapping

If you source file-fit rings, then they must be gapped. Each ring should be gapped in its bore, so they must be marked (number-1, etc.) after gapping. You should gap the ring in the bore at approximately its run height in the cylinder or the second ring is gapped farther down the bore than at the top. Set the depth with a depth mic.

Valve-to-Piston Clearance

One of the most important clearances to check on the piston is the valve-to-piston clearance; and it is especially important on Gen III engines. Typically, this clearance is checked using modeling clay on the top of the piston and below the valves. This requires the cam to be installed and centerlined, the head to be installed, and one cylinder's worth of valvetrain to be installed including valves.

You spray the head's chamber and valves with silicone spray to help keep the clay from sticking to these parts. The head is installed and held on by two bolts: one on either side of the selected chamber. Rotate the crankshaft at least two full revolutions and stop with both valves on the seat. Then remove the head and measure the clearance with a steel scale by sticking it into the clay on top of the piston.

Gen III hydraulic roller cams use a slightly tighter clearance of .080 to .090 inch on the intake and .090 to .100 inch on the exhaust.

Piston-to-Head Clearance

The piston-to-head clearance can be checked in several ways. One is to add an extra lump of clay to the top of the piston outside the chamber area (flat of the piston). Then as you check the valve-to-piston clearance above, you can also check the piston-to-head clearance. You can also (without the clay) bring the piston to TDC and loosen the rod bolts several turns.

Install the dial indicator to the bolt and zero it. Push the piston up to the head and read the indicator for the actual clearance. Perhaps the easiest method is to install the head without the .040-inch-thick gasket and rotate the crank several revolutions. If the piston does not hit the head, the clearance is adequate.

The head gasket is either .027 inch on the 5.7 or .040 inch on the rest of the Gen IIIs, which equals the minimum clearance so you only need .030-inch clearance.

Piston Pins

Although the piston pin doesn't require much oil for lubrication, it does require *some* and that can be an issue. How the pin is retained in the piston or rod can also be an issue. The 6.2 Hellcat engine uses a diamond-coated pin.

Piston Pin Specifications				
	5.7	6.1	6.4	6.2
Pin OD (inch)	.945	.9842	.9451	.9451
Length (inches)	2.475	2.159	2.159	2.066
Weight (pounds)*	158	**	147	***

* The 5.7 engine is supposed to have a pressed pin and the 6.4 is supposed to have a floating pin. The shorter pin is lighter. The pressed-pin rule does not seem to hold true in actual production.
** The 6.1 was supposed to have a 147-grams floating pin, but this could not be confirmed.
*** The 6.2 Hellcat uses diamond-coated pins.

Size and Weight

Generally speaking, you don't have much say about the size of the pin. It is typically supplied with the piston. The Gen III Hemi piston pin weighs about 150 grams while the 5.9 small-block pin from the late 1990s weighs only 5 grams more and it is longer and slightly larger in diameter.

Pressed Pin

A pressed pin is supposed to be used in 5.7 engines, but this rule does not seem to hold true in all engines. For performance applications, I recommend upgrading any pressed-pin setup to a floating-pin package. Remember that both the piston and the rod must be changed to convert the engine.

Floating Pin

Floating pins are listed as the 6.4 setup but also show up in the 5.7. Floating pins are very popular in racing, and they have more clearances to check and more oiling paths to follow. You should check the pin's clearance in the pin bushing in the rod and the pin's clearances in the piston pin towers, one on each end.

The pin in the rod needs to be oiled (pressed pins do not). This is typically done by a small hole drilled in the top end of the rod (I-beam style). In an H-beam rod, two small

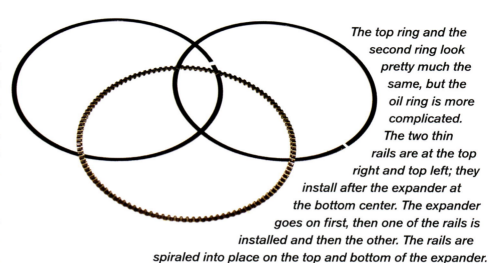

The top ring and the second ring look pretty much the same, but the oil ring is more complicated. The two thin rails are at the top right and top left; they install after the expander at the bottom center. The expander goes on first, then one of the rails is installed and then the other. The rails are spiraled into place on the top and bottom of the expander.

holes are drilled at 5 o'clock and 7 o'clock on the side between the flanges of the H.

Each pin tower should have snap-ring grooves cut into it that receive the pin locks. The pin locks must be used to hold the pin in place. The distance between the two pin locks (once installed, compared to the length of the pin) is the pin's endplay. Different styles of pin locks require different endplay. Typically, the pin should have a minimum chamfer on each end (just enough to remove any edge left from machining).

The Hellcat uses diamond-coated pins, which may make them the best pins, but they are not available individually.

Pin Oiling

The pin in the piston pin tower must receive oil from somewhere. You should always find out how the piston tower is designed to oil the pin and trace the route that the oil travels to be sure there are no blockages.

Pin Locks

Floating pins must have two pin locks: one on each end of the pin. The lock grooves must be cut to match the locks. The piston manufacturer typically supplies the pin locks with the pistons.

Piston Rings

A ring package fits on the piston above the pin bore and below the

Piston Ring End-Gap Guidelines

If you select file-fit rings, you have to decide what gap to file them to. The non-file-fit rings come in .025-inch gaps. The file-fits are much tighter so you can file on the ends of the rings to obtain the gap that you want in your specific bore size. Sometimes these gaps are given as "per inch," but most of the Gen III engines are around 4 inches, so it seemed easy to just list the gap.

These numbers are for the top ring only. In the past, the gap recommendation was a tighter one, but the strat-

Engine	Gap (inch)
Street	up to .025
Drag/oval racing engine	.018 to .020
Nitrous street engine	.020 to .022
Nitrous drag engine	.028 to .030
Supercharged/turbocharged engine	.024 to .026

egy today is to go back to the wide-gap approach for the second ring (same as the top ring, up to perhaps .004 inch larger). ∎

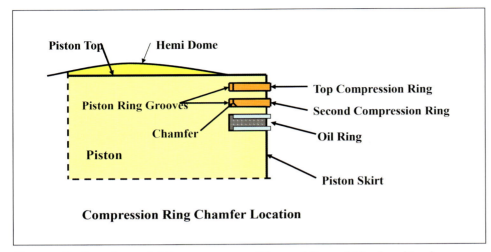

Compression Ring Chamfer Location

The production rings at 1.2 mm are very thin, and it would be hard to draw them to scale and still be able to see them. All three of these rings must fit in the tiny space between the top of the piston pin and the top of the piston. This is the basic ring line-up; it is very short and the rings are very thin.

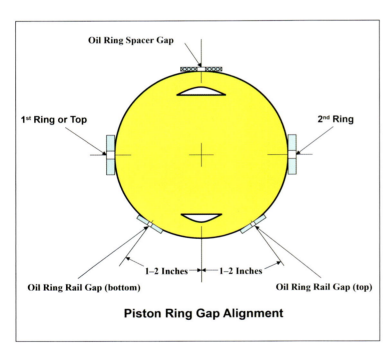

Piston Ring Gap Alignment

It is important that the gaps in the rings are not aligned in any way. As you install various rings onto the piston, gap locations tend to move. That is why it is important to put them opposite from each other, especially the top two rings and the three pieces of the oil ring.

num V-8s and in the 2003–2008 5.7 Hemi. Today the Gen III Hemi (2009 and newer) uses a 1.2-mm (.047 inch) ring package.

Coatings

Most rings today use a moly coating on its face to help seal the combustion chamber and to help reduce friction with the cylinder wall. Use care in handling these rings.

Ring Selection

Never reuse piston rings. With any engine build, new rings are mandatory. The finished bore size in the block and the width of the grooves cut that are in the pistons dictate the rings' measurements. The Gen III engine family uses a 1.5-mm ring package in early engines (2003–2008) and a 1.2-mm ring package in 2009 and newer engines. The 1.2-mm ring is very thin and light. It should be relatively easy to find .020-inch oversized rings for these engines.

Piston Ring Grooves

The top two grooves are generally the same size (thickness). The third ring groove (the bottom one) is much larger (thicker) than the first two; 1.5 mm with a 3-mm oil ring or 2 mm on the 1.2-mm.

Ring Alignment

There are many ways to align the ring gaps of various rings. The real trick is to be sure that they start out not being vertically aligned. Zero alignment. The problem is that the gaps tend to move as you hold the piston and install each ring and this makes it difficult to maintain the proper spacing. By putting them opposite each other initially, they are not lined up and if they move slightly it doesn't align them.

top of the piston. For reference, the standard piston height for the 1993–2003 5.9 small-block piston was about 1.57 inches. Today, the Gen III Hemi typically uses 1.21 inches for the piston height. Everything still fits but the rings have become much thinner and lighter.

The typical ring package includes the top ring, second ring, and oil ring, which comprise three pieces: two skinny (flat) rails and an expender. In the 1970s that basic ring line-up was based on 5/64-inch rings; the high-performance/racing ring was based on the 1/16-inch ring. This evolved to the 1.5-mm (.059 inch) ring pack that was used in some Mag-

LUBRICATION SYSTEM

The oiling system has two basic jobs in a typical V-8 engine: keep oil in the bearings and control windage. The Gen III oil system is fine for mild-performance street builds, but it requires upgrading when elevating into the max-performance territory.

The oiling part of lubrication requires help from almost every part of the engine: block, heads, cam, valvetrain, crank, rods, pistons, and gaskets. The only major parts that are not involved with the oiling system are the intake and exhaust manifolds. The parts that make up the actual oiling system are the oil pump, pan, oil, windage tray, and oil filter.

When upgrading to max-performance, you should install a larger oil pan to increase capacity: a Stage 2 upgrade. A Stage 2 upgrade is also based on the heads and cam. If the bottom end is disassembled, I recommend a Melling pump. However if the engine is not disassembled, I recommend keeping the oil system stock.

A Stage 3 upgrade involves improving volume and pressure. For this level, an upgraded oil pump, such as a high-performance Melling, is desirable. When using forced-air induction, such as a supercharger, you may need to add a high-performance pump.

System Components

The oiling system's main job is to lubricate the engine's moving parts. To do this, it starts at the oil pump, oil pan, and oil pickup. Next is the oil filter mounted on the passenger's side of the block toward the front. The crank mains, rod bearings, and camshaft bearings must have oil. The oil travels to the valvetrain through machined passages in the block and head and is controlled (limited) by the head gaskets. In the head, oil is provided to the two rocker shafts,

Showing the parts of an oiling system is difficult because most of them are inside the block and heads. The oil pump in the Gen III engine fits around the nose of the crank, in front of the timing chain and sprockets. The oil filter fits on the passenger's side.

The stock oil pump fits over the nose of the crank and is driven by the crank through the crank sprocket.

which distribute the oil to each rocker arm and then through the rocker arm to the pushrod and down the pushrod to the tappets.

Drainbacks

The hemispherical combustion chamber layout basically divides the head into two sections. It is very difficult to get the oil on the exhaust side of the head back to the oil pan, so drainbacks are added to. In some racing applications, the drains are added externally, but this approach doesn't work very well on production engines.

Gen III Hemis use four drainbacks per head. There is an oil drainback next to each head bolt except for the front one on each side (five bolts and four drains per head). The front drain omission creates the two cylinder heads (left and right).

Main Bearings

Five main bearings and the number-3 thrust bearing are unique to these engines. The five main bearing shells and bulkheads are also unique. In most earlier small-block and big -block engines, the upper bearing shell (the one in the block) had an oil hole in the center and a 180-degree groove around that shell. The bottom shell was not generally grooved. Gen III Hemi upper shells have a tapered groove for perhaps half of the shell, but it is offset toward one side. There tends to be a second, similar groove (about 1/3 the length of the shell) behind the shell in the bulkhead connected to a second, offset oiling hole in the shell.

Valvetrain

The Gen III Hemi valvetrain functions (oils) similarly to that of the Gen II 426 Hemi. The passage in the head feeds the two rocker shafts and the shafts distribute the oil to the rockers and to the valve tips and pushrod tips. Gen III engines are unique because the oil is fed into the hollow pushrod and down to oil the tappets from the rockers.

Pushrods

Gen III pushrods are hollow, as were the Gen II's, but Gen III pushrods have a pivot on each end, which means that they can be installed in either direction. It also means that the pushrod end of the rocker arm is a cup, with a small hole in the center that allows oil to enter from the rocker arm.

Oil Path

The oil path on Gen III Hemi engines is similar to that of the Gen II in some ways but different in other ways. Both have the oil pump and the oil filter at the front of the engine. However, the Gen III switches the oil filter to the passenger's side of the engine. The Gen III also moves the oil pump inside the pan/crankcase rather than having it on the outside.

After the oil comes back from the oil filter, it travels upward to the main oil galley and then routes downward to all the mains and upward to all the cam bearings. A drilled hole in the block into the main oil galley allows oil up through the block to the head,

The oil pickup attaches to the oil pump and moves the entry to the rear of the pan's sump down the inside of the pan. Trucks generally have a rear sump while the cars typically have front sumps that use different pickups. The VVT versions also use different pickups.

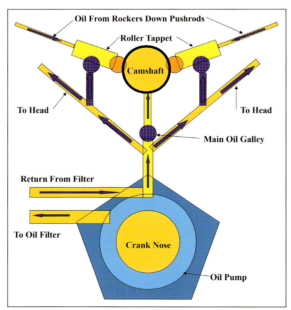

The oil path in Gen III engines is similar to that in the Gen II Hemi: The valvetrain is oiled upward through machined passages in the block and heads and the oil pump and oil filter are at the front of the engine. This is unique because the pump is inside the pan rather than outside the block, and the internal hydraulic tappet is oiled downward from the head through each pushrod.

which is controlled by an orifice in the head gasket.

When it's in the head, the path divides to provide oil to both rocker shafts and to all rocker arms. The rocker arm provides a path for oil to lubricate the valve tips and pushrod tips. The cup in the rocker arm on the pushrod side provides a path for oil to flow into the pushrod and downward to the tappets. The most unique aspect of this oiling system is that the tappet's internal system is oiled from the pushrods and rockers.

The oil returns to the sump of the oil pan, but it is not under pressure whereas the oil path to the heads and to the tappets is under pressure. The oil that leaks out of the bearings (crank and rods) is already in the crankcase. The cam bearings and tappets are directly above the crankcase and have an open pathway to the pan. The oil from the valvetrain (valve, rockers, shafts, etc.) has a special path back to the crankcase via the four drainbacks in the head and matched in the block, which drop the oil down the side of the crankcase and into the pan. Once in the pan, it returns to the sump to be pumped to the engine again.

MDS

Four solenoids located in the tappet chamber cover operate the MDS (Multi-Displacement System). The ECM or computer controls these solenoids. The basic MDS drops four cylinders when the engine is under light throttle. An oil stream to four intake and four exhaust roller tappets pushes a pin in the lifters that allows them to compress rather than move the pushrod, which then drops the cylinder. The tappets can be reactivated within 40 milliseconds if the accelerator is pushed.

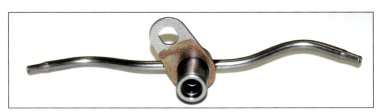

Many Gen III blocks use a special piston oiler or sprayer that sprays oil onto the pistons. It is added to the main oil galley above the mains but below the cam.

Variable Valve Timing

The VVT engine has the peak of the A, or birdhouse, moved forward to add extra oil passages to allow the computer to advance or retard the camshaft's centerline. The cam phaser is mounted to the front of the cam sprocket. The operating solenoid is in the top of the birdhouse.

Leaks

The oil pump is the heart of the lubrication system; maximum pressure and flow occurs at the outlet of the pump. From there basic leakage tends to decrease flow and pressure. Reducing the amount of leakage helps keep more oil in the bearings.

The bearings are designed to leak slightly. Rod side clearance controls the majority of the leakage from the rod bearings, so a high side clearance creates big leaks. The side and roller lubrication of the 16 tappets and the 5 cam bearings are actually small leaks. The rockers, shafts, valves, and pushrods all have small leaks. Therefore, the main player is the crank and rods.

All of these leaks are "designed" leaks that you want to control, not eliminate. As you control the engine's clearances, you also control the amount of leakage.

Crankcase Vacuum

In the normal operation of an engine, a certain amount of gas travels past the combustion sealing system, and that's mainly the rings. In a street car (at least since about 1967), these gases are taken back into the engine by the positive crankcase valve (PCV). For the ring package to work best, you want as low of a pressure in the crankcase as possible, basically a vacuum. To evacuate these gases an evacuation system is added to connect the crankcase to the exhaust header/collector or a vacuum pump can be used.

Crankcase Windage

The design leaks from the five main bearings and the eight rod bearings (as the crank spins at 3,000 or 6,000 rpm) creates a fog of oil,

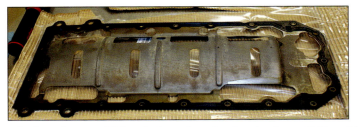

The stock windage tray helps control the oil and helps reduce crank and rod spinning through excess oil, which helps reduce windage losses. This windage tray is standard on all Gen III Hemis, but the 3.58, 3.72, and the 4.00 strokes use different trays. The tray is actually part of the oil pan gasket, rather than being a separate tray. This is a top and bottom gasket seal kit. Production trays use the gaskets and tray bonded together in one piece.

which the crank assembly must spin through. The oil fog creates a drag on the crank that is called windage. The windage creates a loss of power, so you want to control the amount of windage loss by controlling the leaks. This can be accomplished by using a windage tray in the pan, and by using an evacuation system (mainly racing applications).

Oil Pump

The Gen III Hemi oil pump is somewhat unique because it fits around the nose of the crank and is driven by the nose of the crank. Because of this design, they do not use an intermediate shaft. The 2009 and newer oil pump and the 2008 and earlier pump are not the same length. The VVT Hellcat 6.2 engine has a strong, reliable pump. Melling offers a replacement pump and a high-performance upgrade.

The Hellcat comes with a better oil pump, so I recommend using the Melling high-performance pump with any supercharged engine.

Gerotor

Production Gen III Hemi oil pumps are gerotor designs with eight inner lobes and nine outer lobes. The previous small- and big-block (including the Gen II) oil pumps used gerotor pumps with four inner lobes and five outer lobes. The Gen III uses more lobes because it has a larger diameter and because it has the nose of the crank at the center. Gerotor-style pumps are efficient for pumping fluids, including oil.

Rebuild

I do not recommend rebuilding a used oil pump. Rebuild kits can be difficult to find and they are not cheap.

Melling makes replacement oil pumps for Gen III engines and also makes a performance version. (Photo Courtesy Melling)

A pro shop could rebuild the pump, but a new pump may be less expensive. In today's market, any used oil pump that needs to be rebuilt should instead be replaced with a new pump because they are readily available in the aftermarket.

Pressure

The universal guideline for oil pump pressure is 10 psi per 1,000 rpm or 50 psi at 5,000 rpm. The typical stock pump probably puts out about 55 psi based on 5W-30 oil viscosity. The oil's viscosity has

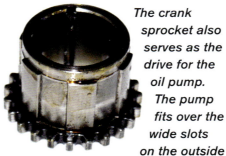

The crank sprocket also serves as the drive for the oil pump. The pump fits over the wide slots on the outside of the sleeve that is part of the sprocket gear. Once installed (the sprocket is keyed to the crank), there is no relative motion between the sleeve and the inner rotor of the pump. It turns with the crank, which creates oil pressure.

This oil pickup fits into the large hole on the right side (the block is upside down). The oil filter installs on the far right side. The oil pickup extends rearward and is attached to the tall stud mounted on the right side of the number-1 main cap. There are two heights. With the aluminum block, you need a special adapter for this pickup-holding feature.

an effect on oil pressure. Engine and bearing clearances have an effect on oil pressure; close clearances help, wide clearances hurt.

Pickups

The Gen III oil pump sits at the front of the engine's crankcase. The oil pump pickup attaches to the pump and extends to the rear edge area of the pan's sump. Therefore, there are at least two basic pickups: the truck type with its rear sump (long pickup) and the car type with the front sump (short pickup). The VVT engines uses pickups that are slightly longer (about 1/2 inch) than the standard pickup because the front of the engine moved forward. If you replace the standard main cap bolts with studs, you need an oil pickup standoff to properly hold the pickup using studs.

Oil Pans

The basic Gen III Hemi comes with one of two unique oil pans. One is the rear-sump truck pan, which is steel, and the other is the front-sump

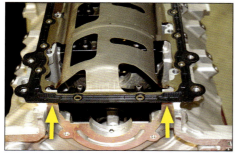

Many aftermarket oil pans add two screws to the oil pan rail (at the rear) to gain improved sealing. They are located about halfway between the last outside screw and the first one on the rear wall; one hole is to the left and one hole is to the right.

Passenger car oil pans are front-sump designs. The front of the pan is to the left here. The passenger car pan is an aluminum casting. The truck pan (not shown) has a rear sump and is made of a steel stamping.

The passenger car pan has several fins and ramps, but their ability to control the oil during acceleration and deceleration seems to be limited.

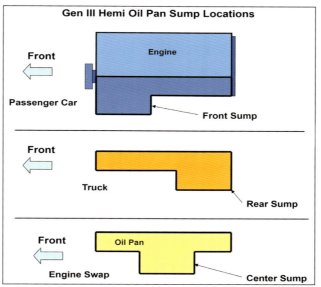

The passenger car uses a front sump and trucks use a rear sump. Most pan manufacturers offer the center sump for muscle car–era vehicles. Because the Gen III Hemi is so popular for swaps into 1964 through 1970 vehicles they made a pan specifically for these applications, which are generally center-sump packages.

passenger car pan, which is made of cast aluminum. Some Jeep models may be in one category and other models in the other category.

Aftermarket manufacturers, including Milodon, Stef, Moroso, and Modern Muscle, offer pans with a center sump for an old chassis, classic car, or old muscle car, which range from 1964 through 1976 and include Chrysler/Mopar A-, B-, E-, and F-Bodies. Typically, these muscle cars have a K-member across the front and a steering link across the rear, which dictates a center sump. Aftermarket oil pans tend to add two screws to the pan rail for better sealing.

The Sump

The sump is the bottom of the oil pan and this is where the oil is collected after it has lubricated the engine. It looks like a square box that has been added to the bottom of the pan (in the front, rear, or center) depending on the model. All production Gen III Hemis use wet-sump oiling systems; a front sump is used in cars and a rear sump is used in trucks. The sump is nearly square and about the width of the block minus the two rows of bolts.

Today, with the new cars and trucks, making the bottom section of the pan wider, generally called a side-bucket or a double side-bucket pan, fulfills the desire for increased oil capacity in the sump. This allows the sump to hold more oil without affecting ground clearance. Caution: With side-buckets, you must be careful of header and exhaust clearances.

Capacity

For the high-performance street engine, since the stock oil pump is pretty good and they all have windage trays, the best performance tip for Gen III engines is to increase the oil pan's sump capacity. The standard setup was designed at 7 quarts, which is much better than many of the earlier 5- and 6-quart pans used on small- and big-blocks.

The main reason for more oil capacity in the sump is that in high-performance activities, the oil has trouble traveling to the sump because it is simply under gravity. In addition, performance activities tend to create accelerations in various directions; it causes the oil to go away from the sump and/or away from the pickup. The easiest way to fix these problems is to start with more oil.

With increased capacity you may want to add baffles around the sump to help keep the oil in the sump and near the pickup. Anytime that you add a baffle to the oil pan, be sure that the oil pump's pickup fits under it and that you can still install the pan.

Ground Clearance

Production cars and trucks are designed so that the oil pan's sump is not the lowest part of the vehicle or engine. Other items that are lower (such as the exhaust pipe) protect the pan from harm on the street. If the sump is dropped, it is pretty sure that the sump will be the first to hit a road obstacle.

For Gen III engines, both Milodon and Modern Muscle offer oil pans with a side bucket. This is added to either side of the oil pan's sump, which makes it wider than the standard sump. This bucket only extends about halfway up the side of the sump. This increases the pan and sump capacity without changing the ground clearance.

For Gen III Jeep packages, a manufacturer such as Modern Muscle could weld a protection plate to the bottom of the sump to help protect the sump in off-road maneuvers.

Baffles

Both Gen III Hemi production pans and aftermarket pans have baf-

The trick with baffles added to the inside of the pan, typically around the sump, is to control the oil so that the crank doesn't have to rotate through it, increasing friction and windage. This Milodon Hemi pan has baffles added to several sides around the top of the sump opening. This pan is designed for older muscle cars, A-, B-, or E-Bodies.

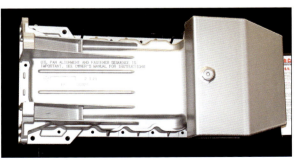

This stock cast-aluminum passenger car pan has had a wide bucket added to the bottom of the pan replacing the stock sump (right). The stock sump is cut off and a sump that is wider than the pan itself replaces it. It sits far enough below the pan rail so that the pan screws can be installed. This design increases the pan's capacity without losing ground clearance. Modern Muscle offers this modified pan and Milodon offers a side-bucket pan.

This all-aluminum pan has a rear sump and baffles inside the sump. It is difficult to see the deceleration baffle, but the Accel baffle is added across the rear of the sump about 2 inches below the pan rail.

fles added. The baffles keep the oil in the sump as the vehicle accelerates, decelerates, or turns. For acceleration, the baffle spans the rear of the sump. Remember that the oil pickup must fit under it and still allow you to install the pan.

The deceleration baffle (braking) is added across the front of the sump at the same height as the basic pan non-sump area. The deceleration baffle is shorter than the rear baffle. If the pan has a rear sump, the rear baffle should be as close to the crank and windage tray as possible. If the sump is in the front or middle, the baffle is at the top of the sump, at the same height as the rear section of the pan.

Windage Tray

The windage tray was created for early-1960s Max Wedge engines and the original 1964 Gen II 426 Hemi, which used two "cork" gaskets: one on the top and one on the bottom. Gen III Hemi trays are actually part of the oil pan gasket. The gasket material is sprayed onto the top and bottom surfaces of the tray, one easy-to-install piece. The production tray must be modified for clearance when using the 4.00-inch- and longer-stroke cranks (see Chapter 3 for details).

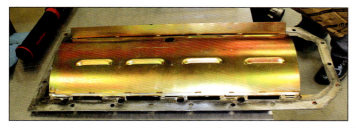

In the aftermarket, windage trays come in many sizes and shapes. Aftermarket trays are designed for use in a specific oil pan. Production trays are designed to fit all production oil pans. This Milodon windage tray is designed to work with the Milodon pan used on an aluminum 426 crate engine.

The basic windage tray is designed to control the oil in the pan and sump during acceleration and braking. It also helps prevent windage losses from the spinning crankshaft. To best control windage losses in the engine, the tray needs to be near the crank but cannot hit it. Therefore most of the windage trays are used with specific cranks: one for 3.58-inch (5.7 and 6.1) strokes, one for 3.72-inch strokes (6.4/392), and another for the 4.00-inch-stroke 426 crate engine and Drag Pak (made by Milodon).

Front Cover

The front cover is cast aluminum and basically provides the front seal for the oil pan and the rear face for the water pump (see Chapter 2 for details).

Oil Filter

The oil filter sits at the front passenger-side of the engine block. In truck and car applications, the filter points straight downward. Some Jeep models change this position and angle the filter rearward at 45 degrees using an adapter to gain clearance to chassis items.

Oiling System Upgrades

The basic upgrade level is determined by the other hardware used in the engine. Refer to "Performance Packages" in Chapter 6 for more information.

Oil

In general, the cycle for oil changes is increasing. In the past, the change cycle was around 3,000 miles, but it is much higher today, at about 10,000 miles on many models. This mileage between oil changes will likely increase in the future because of engine design, assembly improvements, and oil technology gains.

Break-In Oil

As you assemble the engine, when you install the camshaft into the block, and when you install the tappets onto the cam and into the tappet bores, you usually use an engine break-in oil or cam break-in oil. Chrysler/Mopar used Lubrizol oil for this purpose. Crane and Comp Cams have their own cam break-in lubes. In some cases, the cam manufacturer recommends a cam paste. Any cam break-in oil requires that the oil be changed after a short break-in period.

Oil Viscosity

The oil viscosity recommended for the naturally aspirated Gen III engines is a 5W-30 GF-3. The 10W-30 and 5W-40 are higher viscosity oils. The higher viscosity oil tends to make more oil pressure, but it increases the engine's friction and increases

Oiling System Upgrades			
Package	Oil Pump	Oil Pan	Windage Tray
5.7 and 6.1 Stage 1, 2	Stock	Stock	Yes, Stock
5.7 and 6.1 Stage 3	Melling (optional)	Street*	Stock
5.7 and 6.1 Stage 4	Melling High Performance	Race***	**
6.4 Stage 5, 6	Stock	Stock	Yes, Stock
6.4 Stage 7	Melling (optional)	Street*	Stock
6.4 Stage 8	Melling High Performance	Race***	No Space**
All Stage 9	Melling High Performance	Street*	**

* Milodon, Stef, and Modern Muscle street pan.
** Milodon, Stef, and Modern Muscle make trays to work with their pans.
*** Milodon, Stef, and Modern Muscle race pan.

- Stages 1 and 2 are high-performance stock, cat-back exhaust, mild cam, ported throttle body, maybe bracket valve job, etc.
- Stage 3 is high-performance cam, heads, throttle body etc.; Melling pump is optional
- Stage 4 is ported heads (optional), dual purpose and/or race cam, big throttle body, high-rise single-plane intake, headers
- Stages 5 and 6 are high-performance stock, cat-back, exhaust, mild cam, ported throttle body, maybe bracket valve job, etc.
- Stage 7 is high-performance cam, heads, throttle body, etc.; a Melling pump is optional
- Stage 8 is ported heads (optional), dual-purpose and/or race cam, big throttle body, high-rise single-plane intake, headers
- Stage 9 is all engines with added supercharger/turbocharger kits; not listed in package options

the windage. If you have any doubt about which you should choose, refer to your owner's manual.

ZDDP Oil

For older engines, both small- and big-blocks, there have been cam and tappet scuffing issues in the last few years. The general recommendation is that these engines use an oil with the additive zinc dialkyldithiophosphate (ZDDP). All Gen III Hemis use hydraulic roller cams, so this should not be an issue.

Dry Sump

Generally, I do not discuss dry-sump packages for street-strip applications because dry sumps are popular in racing. With a dry-sump system, you replace the oil pump in the engine and the oil pan with new hardware and add an external pump and an oil tank or reservoir. In the past my recommendation on dry-sump systems has been to select one supplier for all hardware and follow the recommendations for your specific application. This is still a good strategy, but there is an exception.

For 2017, Arrow Racing Engines and the Canadian circle-track Pinty's racing series have defined a Gen III Hemi circle-track engine that uses a specific dry-sump system. It has been run on the dyno and at the track successfully. Stef's makes the steel pan and Dailey Engineering makes the three-stage pump. Both the steel pan and three-stage pump are mandated. This system might be a good one to copy.

The race sanctioning body determines the legal three-stage dry-sump pump. Without specific rules, the more typical dry-sump pump has

This three-stage dry-sump pump by Dailey has two suction stages and one pressure stage. This is somewhat uncommon because most dry-sump systems use many more stages; five is common. The stages are stacked together and held together by long, thin bolts. Cog belts off the front of the crank drive these pumps.

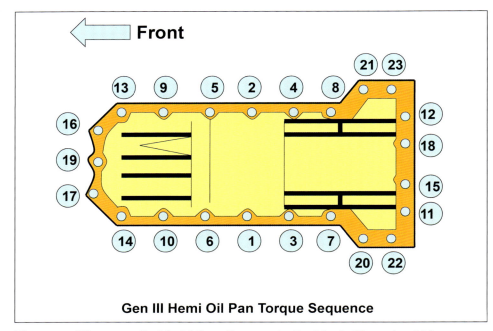

Gen III Hemi Oil Pan Torque Sequence

There are 23 screws that hold the oil pan onto the block. They should be torqued in sequence to 106 in-lbs.

four or five stages with the extra stages added to the oil pan or crankcase. Manufacturers include Stef's, Milodon, Dailey Engineering, and Arrow Racing Engines.

Vacuum Pump

The basic vacuum pump is designed to remove blow-by and other contaminants from the engine's crankcase. Blow-by is the combustion gas that travels by the rings during normal engine operation. The two levels are an evac system using the exhaust (see Chapter 12) and an actual vacuum pump.

Moroso is the leading manufacturer of vacuum pumps for this application. The vacuum is generally used in racing, not in dual-purpose applications. These vacuum pumps can create about 15 to 18 psi vacuum in the crankcase, and this amount of vacuum can offer up to 2 percent power gain in the 5,000- to 7,000-rpm range.

CAMSHAFT, LIFTERS AND CAM DRIVE

The camshaft used on Gen III Hemis is typical of most modern V-8s, with five cam bearings and four cam lobes between each set of cam bearings. If the cam is out of the engine, the most notable aspect of the Gen III is the size of the cam bearings. They are visibly larger in diameter, especially the number-1 journal on the 2009 and newer VVT engines, when compared to older V-8s, small- and big-block. Another interesting feature is that there is no gear on either end of the cam because there is no intermediate shaft. Interestingly the number-5 bearing is small, similar to the Mopar small-block's number-5.

In this chapter, I provide guidelines for selecting a new camshaft for your Gen III engine. You have many choices, but the selection process is somewhat easier because you only have one group of cams, hydraulic rollers. Since all production engines use the hydraulic roller design, it makes no sense to back up to an older hydraulic flat-tappet or a mechanical flat-tappet.

Camshafts

All Gen III cams are hydraulic roller designs. They have very large main journals and no gear. VVT versions have a wide front of number-1 cam bearing/journal. Other than the large diameter allowed by the big cam bearings, the hydraulic roller cam doesn't look any different from other styles of cam.

Identification

Gen III Hemi engines do not have a distributor drive gear at the front or back of the shaft. In addition, a small number-5 bearing journal is at the rear of the cam. The other three journals are basically the same size (around 2.25 inches) and the front journal is very large at 2.67-inch on VVT versions. VVT cams (2009 and newer) also have a wide number-1 journal. Cams are made of steel or very hard cast iron.

Most production cams have a three-digit code plus two letters such as "AA" (the last three digits and two letters of the part number) stamped

VVT cams (top) have two grooves in the number-1 journal along with the extra length (about .460 inch) of the number-1 cam journal. Note the small holes drilled into the bottom of the grooves. That is all part of the VVT system.

Most camshafts look alike. More lift and more duration in the lobe profile is not generally visible. All Gen III engines use hydraulic roller cams. There are two basic groups: standard and VVT. The top cam here is a VVT. Look at the multi-grooves in the number-1 journal (far left).

5.7 Hemi Camshaft Options			
Engine	Model	Part Number	Options
5.7	Challenger MT	53022064BD	VVT
5.7	2500 Ram	53022314AD	VVT and non-MDS
5.7	1500 Ram	53022263AF	MDS, VVT, and active intake
5.7	Jeep and Durango	53022372AA	VVT and MDS, 2011-newer
5.7	Aspen and Durango	53022065BE	Hybrids, 2009

At the number-5 journal face, there are three numbers and two stamped letters to help identify these production cams. Although the lift and duration change with displacement, the actual cam part number changes for various technologies, such as MDS, VVT, active intake, and hybrid. This example is marked "379," which is the last three digits of its part number; the two letters are "BC." It is not a 5.7 cam but one of the high-lift 6.4s.

Most older small- and big-block V-8s had number-1 journals around 2 inches in outside diameter. Gen III engines use a much larger number-1 2.67-inch journal, which is partly related to VVT.

on the face, along with another stamped number, five or six digits, which is an internal number but doesn't help with identification. A lot of the earlier in-plant identification was done by small paint stripes on the cam core, which tend to fade with mileage (heat) and become unreliable for identifying cams in used engines. Most aftermarket cams are stamped with profile numbers and/or manufacturer numbers.

Currently, there are two 5.7 cams and one each for the other three engines, or five total based on the lobe profile, lift, and duration. The most popular engine, the 5.7, since about 2009, has had five different cams based on part numbers, which all have the *same* lift and duration.

A good possibility exists that many of these cam part numbers will change to one or additional numbers. Therefore, the current numbers may be superseded, replaced, or updated in the future. The stamped cam number is the last three digits plus the two letters. From a performance standpoint, cams are mostly made for two lifts:

.475 and .570 inch. These two-valve lift numbers are very important to the cam selection process.

Journals

Small- and big-blocks typically have the cam journal diameters decreasing in size by .001 or .002 inch per journal through the first four and then dropping to 1.72 for number-5. Big-blocks, including Gen II versions, continued the sequence to the number-5. All of these cam bearings started the sequence at just under 2.00 inches.

Function

In any engine, the cam's main job (in conjunction with the valvetrain) is to open the valves in the head. That sounds easy, but the field is full of potholes. The cam lobe translates its movement to the tappets and the challenge is to transfer this movement to the valves in the head maintaining a specific relationship to the crank and pistons. As with earlier big- and small-block engines, Gen III Hemis are four-cycle engines (intake, compression, power, and exhaust, or ICPE).

Gen III engines feature hemispherical-style combustion chambers in the cylinder heads, which means that the engines are overhead valves (OHVs) with one camshaft in the center and with two valves per cylinder, one on each side of the chamber. Two valves per cylinder translates to two lobes on the

Common Camshaft Journal Specifications			
		5.7 and 6.1	6.4 and 6.2
Bearing Journal Diameter (inches)	number-1	2.29	2.67
	number-2	2.28	2.28
	number-3	2.26	2.26
	number-4	2.24	2.24
	number-5	1.72	1.72

CAMSHAFT, LIFTERS AND CAM DRIVE

Just to the left of this number-5 journal is a small lobe on the top cam but not on the lower cam. This is the MDS lobe. No MDS? No lobe (lower cam). If the solenoids are disconnected, it has no affect on engine operation.

Cam Specifications					
	5.7	5.7 Eagle	6.1	6.4	6.2
Lift (inch; intake/exhaust)	.472/.460	.472/.460	.571/.551	.571/.536	.571/.536
Duration (degrees; intake/exhaust)	260/268	258/288	283/286	286/288	278/304
Overlap (degrees)	37	39.5	50	46	51
Centerline cam (degrees)	112	114	109	113	109
Centerline installed (degrees)	110	114	108	115	117
Type	Hydraulic Roller	Hydraulic Roller	Hydraulic Roller	Hydraulic Roller	Hydraulic Roller

camshaft for each cylinder. The hemi configuration means two rocker shafts, such as an intake shaft and an exhaust shaft.

The cam lobes on the camshaft must line up with the tappet bores in the block, 16 total. Gen III Hemi engines use hydraulic roller cams and tappets, so a yoke or guide is required to keep the tappets straight, relative to the cam lobe, and thus they don't turn in the tappet bore as they go up and down through their cycle. Gen III production engines have four tappets per guide or yoke, and these guides bolt to the block.

The Gen II Hemi and the A and B/RB engines oiled the valvetrain up through machined holes in the block and head; Magnum small-blocks oiled the valvetrain up through the pushrods from the tappets. Gen III Hemis borrow a little from each: They oil the valves up through machined holes in the block and heads and they oil the tappets (internally) down through the pushrods.

Valvetrain

The Gen III family of engines oil the valvetrain up through machined/

drilled holes in the block and heads and then oils the tappets down through the pushrods from the rockers and shafts. This system seems to be flawless. Racers being racers, they may switch to a standard oil-through-the-pushrod system, which oils up through the pushrod. To date, the hardware to do this is not available, which actually means that it hasn't been developed or offered for sale as a package. It appears to offer no advantage at this time.

Tuning

All Gen III Hemis use MPI; the MPI system uses six or seven sensors and a very sophisticated computer to control both the fuel system and the ignition system. With the MPI system, if you install a camshaft with more lift, more duration, and/or more overlap, the computer must be "re-programmed," or reflashed. The bigger cam tends to make more power and more power requires more fuel. The computer controls the fuel, so without a "re-program" to increase the amount of fuel, the engine runs too lean, and you have problems. In reality, it is more complicated than this simple example, but you get the idea.

Generally, the production computers can't be re-programmed, but the aftermarket has supplemental devices that can be added to solve this problem (see Chapter 10 for more details). Several years ago, SCT re-programmed the Chrysler computers. Today, SCT offers a hand-held device that works with the production ECM. This approach seems easier, less expensive, and faster and many other aftermarket manufacturers have followed suit.

Camshaft Technology

Gen III Hemis are typical of current high-tech OHV engines that operate on the four-cycle process. Each cycle, or phase, has a crankshaft length of 180 degrees. Therefore, the full engine cycle takes two revolutions of the crank. One full revolution is 360 degrees, so the full engine cycle is two full crank revolutions (720 degrees). In these four-cycle engines, the cam runs at half the engine speed, which means that the cam sprocket is twice as large as the crank sprocket. What makes the Gen III unique is that the cam is raised so high in the block that the chain is very long, relative to other OHV engines.

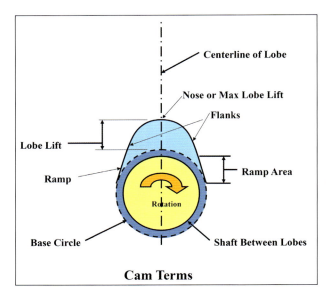

Cam Terms

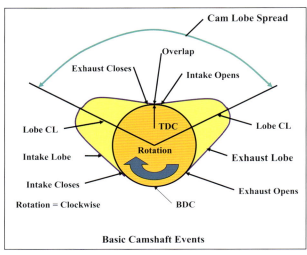

Basic Camshaft Events

One feature of the Gen III engine cam is its bigger base circle and larger cam journals. The large journals allow the bigger base circle cams because the lobe must fit through the cam journal. This allows bigger cams or the same cam (lift-wise) will be stiffer and stronger.

The cam manufacturer sets the lobe centerline angle (LCA), and it places the intake and exhaust lobes on the cam blank. Once ground, you can install the cam relative to the crank (the installed centerline) in many positions. VVT uses the phaser on the front of the cam to rotate this whole diagram relative to TDC and BDC.

tion (production) and the duration at 50 (race).

Advertised duration is the most popular for street or dual-purpose cams and probably the most common. Advertised duration is loosely defined as the point at which the tappet is about .005 to .017 inch off the seat (zero position, or base circle), but the specific number varies with the manufacturer.

With duration at 50, the tappet is exactly .050 inch off its zero position (base circle) and is the same for all manufacturers. Duration at 50 is always smaller than advertised duration. However, duration at 50 is much more useful in all-out race cams and engines. The typical OEM manufacturer does not list the production cam's duration at 50.

Events

There are four basic cam events: intake opens, intake closes, exhaust opens, and exhaust closes. These four events are expressed in degrees relative to TDC or BDC. The cam's duration can be calculated if you know these four events. The 5.7 Eagle (2009 and newer) has an intake duration of 258 and an exhaust duration of 288 degrees. With the four events, you can also calculate the cam's centerline. If you know the cam's duration and centerline, you can calculate the four events.

Overlap

A cam's overlap is the amount of time that both the intake and exhaust valves are open. It is the sum of the intake open event (before TDC) and the exhaust closes event (after TDC). Street cams have low overlaps and race cams have high overlaps. However, street cams use advertised duration numbers, which

Lift

The cams' lift, or lobe lift, is possibly the most easily observed specification on a cam because the lobe becomes taller as the lobe lift increases, but the actual height difference might only be .060 inch at the valve of .100 inch (.571 versus .475 inch). By definition, the cam/lobe lift is the distance that the tappet moves upward in the tappet bore or the difference in height between the nose of the cam (max lift) and the base circle (zero lift).

Valve lift (the distance that the valve moves) is cam lift multiplied by rocker ratio. Two lifts are available for production Gen III Hemi engines: .475 and .570 inch. There are also two basic rocker ratios: the standard 1.65 and the Drag Pak 1.85. Valve lift is easiest cam specification to directly measure on an assembled short-block.

Duration

Another cam spec is the cam's duration. It is defined as the number of crank degrees that the intake and exhaust valves are open, or off their seats, in the head. A cam has two types of duration: advertised dura-

Cam Specification Conversion

When you buy a new aftermarket cam, you also receive specifications and tips for installation. There are two methods for installing a high-performance camshaft: event and centerline.

The aftermarket favors the event method while the factory favors the centerline method. Both begin by lining up the dots, which gets you in the area, but you want to be at one specific location not plus-or-minus 8 or 10 degrees.

With the centerline method, you are provided the profile's duration (advertised) and the installation centerline. With the event method (popular aftermarket approach) you have five numbers on your installation sheet: the profile's duration and the four events. To convert these five numbers to the centerline method, perform the following steps (I just look at the intake side here).

Let's assume that your new cam has an intake duration of 258 degrees (5.7 Eagle, for example) and the intake opens at 15 degrees before TDC and closes at 63 degrees after BDC. This is typical event method data. First you add the numbers (15 + 63 + 180 = 258). Next you divide the duration in half (258 ÷ 2 = 129). Since the intake opens at 15 degrees before TDC, the center has to be located 129 degrees later, or at 114 degrees (129 - 15 = 114).

On the other hand, you have two numbers on your installation sheet: duration and centerline. Let's now assume that the intake has 286 degrees of intake duration (on a 6.4, for example) and an intake centerline of 113 degrees. First you divide the duration in half (286 ÷ 2 = 143). Now you know that the intake opens at 30 degrees before TDC (143 - 113 = 30) and the intake closes at 76 after BDC (30 + 180 = 210) and (286 - 210 = 76). ∎

yields an overlap number in the 25 to 50 or 75 area. Using the duration at 50 numbers for these cams ends up yielding very low or zero overlap numbers, which isn't very useful.

Velocity and Acceleration

Cam lobe velocity is defined as how fast the lift changes. Acceleration is defined as how fast the velocity changes. Although both are important to the cam profile and its performance in the engine, these are aspects that are designed into the cam profile by the manufacturer.

Valvetrain Geometry

The geometry, or alignment, of the valvetrain starts with the camshaft location and the tappet angle. All production Gen III Hemi blocks use the 75-degree tappet angle, up from the small-block's 59 and the Gen II's 45. Because the cylinder head design is hemispherical with one valve on each side of the chamber, unlike the wedge's side-by-side

configuration, the tappet angle is always a compromise between the intake and exhaust sides. So far, this stock tappet angle appears to work best.

Potential Benefits

The 6.4 production cam has a .571-inch lift and long in duration at 286 degrees. More valve lift and longer durations make more power. Move the velocity and acceleration numbers in the design equation to increase the lift. You can also change the cam's centerline (ground), which allows you to adjust the cam's overlap.

One of the benefits of the Gen III Hemi's larger cam bearing diameters is that it makes the shaft stiffer. This stiffness reduces lobe deflections due to high valvespring loads, and that comes from better valvesprings and more lift.

For example, the nose of the cam profile must fit through the cam bearing (2.24 inches is the smallest), and it's about 1/4 inch bigger than

earlier V-8's, which are less than 2.00 inches. The 6.4 cam with .571 valve lift needs about .356 inch of cam lift, or distance from the nose to the base circle. The Gen III has a .120-inch advantage. Going to a .675-inch lift cam, uses up .066 inch of the advantage, but the .675-inch is the biggest cam made for the Gen III.

The shaft of the cam must be lower than the base circle to allow the profile to be ground. As the cam lift increases, the base circle becomes closer to the shaft; too close and you need a new casting or the cams must be ground from billets (more expensive).

Installation Basics

There is very little that you do to the camshaft itself. Most installation tips and tricks relate to how it is installed into the engine. It is the one thing relative to the cam that you change as you build the engine. The first step: When removing the timing chain at disassembly, retract the

Camshaft Terminology

The main camshaft specification terms are lift, duration, and overlap. Camshaft events are closely related to the valve events and the basic specifications. You need a working knowledge of these terms to be successful at making your cam selection.

Lobe Centers

Sometimes called lobe separation angle (LSA), this specification is ground into the camshaft by the cam manufacturer and is the angle of the intake and exhaust lobes relative to TDC. In general a low LSA (less than 110 degrees) produces better peak horsepower; a wider LSA (more than 112 degrees) helps torque and spreads the power over a greater range of RPM.

Once the cam is ground by the manufacturer, a unique feature of the cam is that as you move the intake lobe in one direction, the exhaust lobe moves in the opposite direction. For a 6.4 cam, for example, the intake centerline is 113 degrees and the exhaust is 115 degrees. If the cam is advanced 4 degrees, the intake centerline moves to 109 degrees and the exhaust moves to 119 degrees.

VVT

The VVT system changes the cam's installed centerline while the engine is running. The phaser, which is bolted to the front of the cam sprocket, accomplishes this change (see Chapter 3 for more information about VVT).

The cam phaser lock on the lower right is designed for VVT cams to lock it into one position. (Photo Courtesy Comp Cams)

Base Circle

The base circle is the lowest part of the cam lobe, or profile, closest to the center of the camshaft. It is also where the tappet spends most of its time during the full engine cycle. As the cam lobe has more cam lift, the base circle of the lobe is ground

closer to the center of the shaft. You can't add lift to the nose because the cam would not fit through the cam bearing shells.

Nose

The nose of the cam lobe is opposite the base circle, the highest point of the lobe, farthest from the center of the camshaft. The nose has a rounded point; the longer the duration of the cam, the more rounded the point becomes. For any given engine, the nose is almost always at the same height relative to the center of the cam. This is because the nose must fit through the cam bearings, which remain at the same diameter.

Cam lift is gained by machining down the base circle. The exception to this is if you can go to larger cam bearings (in diameter), which allows the nose height to increase and the shaft becomes thicker and stiffer. Obviously the bigger cam requires a new casting be made or the cam must be made from billet.

Profile

Everything between the base circle and the nose on both sides, opening and closing, is considered the lobe's profile. A profile is generated by a long mathematical equation and

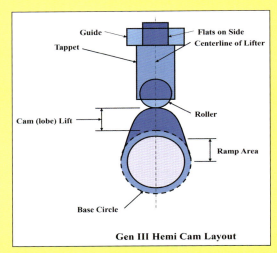

Gen III Hemi Cam Layout

Ramps are part of every cam lobe profile and the opening ramp and closing ramp are not the same. The ramps may be similar, but they are not exactly the same. The opening ramp's job is to get the tappet moving upward, away from the base circle. On the other side, the closing ramp's job is to gently allow the tappet to come to rest on the base circle. The two ramps are the key to the smooth operation of the cam. This aspect is important in production cams and all street engines. The guide at the top fits over the flats on the sides of the tappet. The guide keeps it square to the lobe, so this is critical with any roller cam.

run (calculated) in computers. Every manufacturer has its own equations, which are closely guarded secrets. The results have a number with lift and duration specs.

Lobe Taper

Lobe taper is the angle at which the lobe is ground relative to the centerline of the camshaft. With roller camshafts, there is no taper and all Gen III engines use roller camshafts.

Valve Lash

Valve lash is used with mechanical cams only. It is defined as the free lash, or gap, between the rocker arm and the valve tip when the tappet is on the base circle. The typical valve lash for a mechanical roller camshaft is between .020 and .035 inch. The ramp design used by the manufacturer has a lot to do with the valve lash recommended by the manufacturer. ■

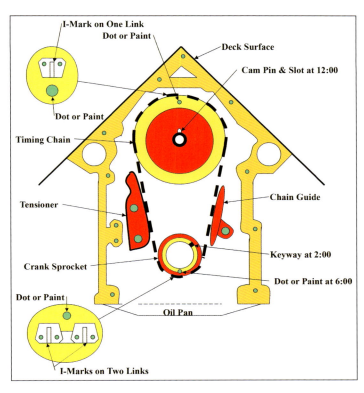

Aligning the dots on the Gen III engine family is a little more complicated than it was on the older engines. The key is finding the plated links (term used by service: they look more like an "I" marking to me). There is one by itself (top) and then two that are next to each other (bottom).

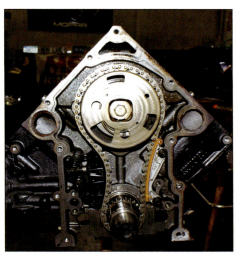

It is much farther between the crank sprocket and the cam sprocket than on other engines. This allows room for the tensioner (lower left). Chain removal starts with the tensioner.

tensioner shoe (left side of the chain as you face the engine) until the hole in the shoe lines up with the hole in the bracket. Then insert a pin to hold it in place. Finally, remove the cam sprocket, chain, etc.

Aligning the timing marks, also called lining-up-the-dots, each tooth in the cam sprocket is about 7 to 7-1/2 degrees. So aligning the dots is only accurate to about 7 degrees. The installed centerline could be 107 or 121, a tooth off in each direction. If you want to install the cam at 112

or 118, you can't get there with the lining-up-the-dots method.

Performing the lining-up-the-dots method as the first step in the cam installation process will save you a lot of time in the long run. The dial indicator also catches a one-tooth-off error. The Chrysler service manual describes the aligning process as follows:

First, install the tensioner assembly and the torque mounting bolts to 250 in-lbs.

Second, retract the tensioner.

Third, the camshaft pin and the slot in the cam sprocket must be clocked at 12:00.

Fourth, the crankshaft keyway must be clocked at 2:00. The crankshaft sprocket must be installed so that the dot paint marking is at 6:00.

Fifth, the production timing chain must be installed with the single-plated link aligned with the dot paint marking on the camshaft sprocket (at 12:00).

Sixth, the crank sprocket is aligned with the dot paint marking on the sprocket between two plated timing chain links (at 6:00).

The chain tensioner is spring loaded. It must be pried toward the left until the hole in the tensioner lines up with the hole in the attaching bracket. Then slide a pin into place so the tensioner stays retracted and you can remove the chain.

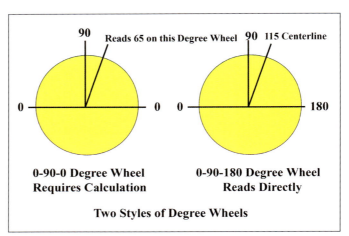

0-90-0 Degree Wheel
Requires Calculation

0-90-180 Degree Wheel
Reads Directly

Two Styles of Degree Wheels

The degree wheel is a thin, flat, and round plate that has markings around the outside edge that indicate degrees of a circle. High school geometry says that a circle is divided into 360 degrees. Engine builders use a degree wheel to accurately install and centerline (degree) camshafts and to check camshaft locations relative to the crank. There are two styles of degree wheels: the 0-90-0 layout and the 0-90-180 layout. Caution: On the 0-90-0-style wheel, the typical cam centerline for a Gen III engine is 115 degrees, but it does not read directly.

Seventh, place both sprockets inside the chain and install the sprockets over the shafts and double-check alignment marks.

Last, install the camshaft bolt (90 ft-lbs).

Cam Degreeing

The centerline installation method or process is often called degreeing the cam, which is accurate because the degree wheel is used for all of your readings and those readings are in degrees. The event method of installation also is called degreeing and also uses a degree wheel.

Always remember that what the cam wants is the specific relationship with the number-1 piston so the zeroing process for the degree wheel and the dial indicator, which is the first step, is very, very important.

Advancing and Retarding

The process of degreeing a cam, that is, the cam movement (changing the centerline) is called advancing or retarding the cam. For example, if you start with a 115-degree centerline and move the cam centerline (intake valve) toward 100 degrees, it is consid-ered advancing the cam. If you move the centerline away from 100 degrees, it is considered retarding the cam.

Zeroing the Degree Wheel

The first step in the process of degreeing the cam is to zero the degree wheel to the top of the number-1 piston. Rotate the crank until the number-1 piston is at TDC. Install the dial indicator perpendicular to the flat of the piston. Install the degree wheel on the nose of the crank and then install an adjustable (perhaps a bent coat-hanger wire) pointer. Make it point to zero on the degree wheel. Rotate the piston until the indicator says that it is at the maximum vertical travel; zero the indicator. Rotate the crank clockwise until the indicator reads exactly .050 inch down. Stop and read the degree wheel and write down the measurement.

Now rotate the crank in the counterclockwise direction, through zero and down to exactly .050 inch on the other side. Read the degree wheel again. If the pointer zero is correct, the two readings should be the same. If the first reading is 20 degrees and the second reading is 28 degrees, the pointer is 4 degrees off (28 - 20 ÷ 2 = 4). Move the point from 28 degrees to 24 degrees and repeat.

Centerlining

The process used to accurately install a cam or degree a cam is called centerlining. With the cam installed in the block, the next step is to line up the dots on the crank and cam sprockets, and install it onto the cam and crank. The important thing is to position the number-1 piston at TDC and line up the dots.

Lining up the dots, the keyway, and the marked links is the first step in installing the cam correctly. It will save time and effort in the long run. It can help to fine-tune this process if you use a steel scale to line up the dots with the centers of the crank and cam sprocket.

Install the two tappets into the tappet bores for the number-1 cylinder (an intake and an exhaust). Lube with very thin oil. Reposition the dial indicator from the top of the piston to the top of the intake tappet. It should be on the tappet's outside edge, which has a flat surface, and it should be parallel to the tappet centerline.

Rotate the crank one full revolution and return to TDC with both tappets on the base circle. Carefully rotate the crank in the clockwise direction until the intake tappet reaches max lift and zero the indicator. Rotate the crank clockwise until the tappet reaches the base circle and starts to lift upward again. Stop at .050 inch below max lift and read the degree wheel (90 degrees, for example). Continue rotating clockwise through max lift and back to the .050-inch down point and take a second reading (140 degrees, for example). The centerline is halfway between these two points (90 + 140 = 230 ÷ 2 or 115 degrees. Actual readings (90 and 140, for example) will vary with cam.

In this example, if you wanted the cam installed at 115, you are done. If it measures 107 degrees, you are a tooth off on the sprockets or the dots are not lined up exactly by one tooth. If you wanted to install the cam at something between these two numbers, you need a multi-keyed sprocket set or similar device.

Centerlining Hardware

The Gen III Hemi is new to the performance hardware business and the specifics are still evolving. A small pin in the nose of the cam controls the cam centerline: it fits into a matching hole in the cam sprocket similar to Gen II Hemi engines. The problem is that the pin in the Gen III engines is smaller, and bushings are not available to match the size of the pin.

The degreeing procedure in the Gen II used small offset bushings. These bushing are readily available at any speed shop, but the hole is too large for a Gen III. At this writ-

Once the camshaft has been degreed, a thrust plate secures the cam in the block. The early 5.7 and 6.1 engines use a small plate that bolts to the front of the block. Newer VVT engines use a much larger thrust plate. The aluminum block uses the early-style small thrust plate. Four bolts hold the plate in place, and the large hole at the top is for the VVT solenoid.

ing, bushings with the smaller hole are not available. A multi-keyway sprocket can be used on the early 5.7 (2003–2008), but is not available yet for the newer engines. Adjustable sprocket sets are used on some engines but haven't been developed for the Gen III yet. You might try drilling out the dowel hole in the front face of the cam down to the Gen II size. Then install the Gen II pin and use the readily available bushings.

Pushrod Length

Pushrods are actually part of the valvetrain, but when Chrysler engineers revised the valvetrain to work with the high lift (.571 inch) used on 6.4 engines, the rocker shafts were moved upward to accommodate the taller installed height (2.00 to 2.050 inches) and that improved geometry. This taller arrangement requires lon-

ger pushrods on both the intake and exhaust sides.

Valve Clearance

The valve-to-piston clearance is important and with bigger cams, valve notches typically must be cut into the top of the piston. In general, advancing the cam (moving centerline from 115 to 105 degrees) hurts valve-to-piston clearance; retarding the cam (moving the centerline from 110 to 118 degrees) helps valve-to-piston clearance. Therefore, you should measure the piston's valve-to-piston clearance early in the assembly process with the cam properly centerlined to avoid future problems. This way you can cut the notches (if required) into the pistons early on.

The Gen III family of engines tends to be tight on valve-to-piston clearance because it has a fairly high production compression ratio (about 10.5) and relatively high valve lift (about .480 to .570 inch).

Operation

If you install a bigger cam into an MPI engine, it may require the ECM to be reprogrammed. On engines with VVT, the typical big-cam-lope may not occur. There are several reasons for this, but the ECM is adjusting the centerline to effectively lower the cam overlap, which limits the observed lope commonly caused by higher overlap. This could be addressed by programming or by slight adjustments to the cam. Currently, these cams are in development by Modern Muscle.

Cam Bearings

The five cam bearing shells are pressed into the block. Gen III engines use much larger diameter

The cam-bearing bore is at the top and the main bearing bore is at the bottom (of similar size).

Here is a timing chain and crank sprocket (bottom) and cam sprocket (top) with the VVT phaser. Both versions of the standard timing chain are silent. The trick is to find the plated links. The term "plated" is the service word. It looks as if it has a laser-etched "I" between the two pivots (I-marked). The crank sprocket (about 2 inches thick) and the wide ridge on the outside drive the oil pump. The I-marked link is at the bottom, just to the right. There are two different styles of timing chain and sprockets. They must be changed as sets. The oil pump slips over the sleeve on the crank sprocket (left).

cam bearings than earlier V-8 engines, 2.43 versus 2.00 inches. Durabond, Clevite/Mahle, and Hastings offer standard bearings. VVT engines use an extra-wide front cam bearing (number-1) that has two slots in it.

Roller Bearings

There is no reason to install roller cam bearings in a street or dual-purpose engine. Racers use roller cam bearings. The 50-mm roller bearing can be installed, but the camshaft ends up at a 2-inch diameter. Race cams have moved into very high lifts and this diameter is too small. The 60-mm roller cam bearing now used is too large to fit a cast-iron block but does fit an aluminum block. Remember that Gen III engines have larger bearing journals in stock configuration, which allows more cam lift to be ground before you get to the roller bearing cam situation.

Cam Drives

The cam is driven from the crank by a cam drive system. There are several styles and each offers certain advantages. You always want to replace all three pieces of the cam drive if it appears to be showing too much wear. The cam drive should be replaced on high-mileage engines and on any 2003–2008 engine rebuild.

Timing Chain

The production timing chain system was changed from the early engines (2003–2008) to a new design (2009 and newer). The old chain had 86 links and the new chain has 76 links. Obviously, this change in the chain means that the sprockets

The cam phaser, part of the VVT system, is mounted to the cam sprocket via bolts from the opposite side.

are also changed: 26 and 52 on the early units to 23 and 46 on the newer units. The new chain is narrower than the original unit. Both of these designs are considered "silent" chain designs. The new setup seems to be strong; it has been used in racing applications without failure. Manley offers a double-roller setup.

Removing the cam's thrust plate allows you to remove the cam. Newer engines have a much larger plate that works with the tensioner and guide.

On VVT engines (2009 and newer) there is a large, round phaser mounted onto the front face of the cam sprocket. The VVT engine design allows the camshaft centerline to be changed while the engine is running; the phaser accomplishes this feat. The phaser is about 1-inch thick and about 6 inches in diameter. If you do not want this phaser to operate, Comp Cams offers a phaser lock (PN 5760) that locks the cam in place.

Early engines used a small thrust plate (top, upside down) along with the spring-loaded tensioner (lower left) and the guide (lower right). The pin in the tensioner at disassembly holds the tensioner from expanding.

Tensioner and Guide

All Gen III engines use a tensioner and guide system to keep the long chain at the proper tension. I recommend replacing the original 2003–2008 tensioner and guide with a newer unit.

Belt Drives

Belt drives for the cam are generally considered a race part and so far none have been introduced for Gen III engines. It should be easy to install because the tensioner has two bolts that attach it that could easily be used for the idler/tensioner used in belt drives.

Front Cover

The front cover (aluminum casting) covers the cam drive. There is one for standard engines (2003–2008), one for VVT versions (2009 and newer), and one for use with a distributor/carburetor from Arrow Racing. Truck front covers may also be different. (See Chapter 2 for more details.)

The backside of the cam sprocket has five flush, attaching screws that hold the phaser to the cam sprocket.

The special front cover that allows the use of a distributor on Gen III engines uses a drive gear on the nose of the cam. The gear and longer bolt are part of the Arrow Racing front cover package.

Tappets

The hydraulic roller tappets used in Gen III engines are tall. There are two basic styles of hydraulic tappet: the older ones (2003–2008) and the newer VVT lifter. VVT lifters have a chamfer on the top. The VVT lift

The standard hydraulic roller tappet is quite tall. The top section has two flats on one end. Tappets fit into the guide to keep it square to the cam lobe during operation.

The standard non-MDS hydraulic roller tappet has a smooth, solid barrel; no holes in the side of the barrel. The MDS tappet (not shown) has a small hole in the barrel about halfway up.

can be used in the old engines, but the old lifters can't be used in the newer VVT engines. The VVT lifter is designed to work with the high lifts used in the 6.4 (and newer engines). Therefore, the VVT lifter is the best for performance also. All of these tappets receive their internal oil down from the pushrods.

In all MDS engines, there are two styles of lifter; racers tend not to like the MDS lifter even though it works fine in racing applications with the solenoids disconnected. If you

Hydraulic roller tappets are installed in groups of four. These are non-MDS tappets; all four are the same. The MDS system uses two tappets in each guide. The guide has two flat widths to pilot the tops of the tappets so they end up in the right locations. If you use four non-MDS tappets, popular with many racing builders, the guide must be changed to a non-MDS version.

Jesel offers larger tappets with a tab, or foot, at the top. They come with a sleeve that is pressed into the tappet bore after it is sized. The sleeve has a vertical slot that serves as a guide for the tab. It's mainly a race part.

With the head removed, you can see and remove the tappets. They sit in groups of four in the plastic guide. Guides are available for MDS and non-MDS tappets. The bolt that holds the guide in place is in the center. The guide fits over the slots on the top of each tappet.

One set of four tappets removed from the engine.

Jesel makes this metal tappet guide to replace the stock plastic unit.

switch to 16 non-MDS lifters, you must switch the guide bar also.

If you use the engine for performance applications and do not want to use the MDS feature of production Gen III blocks, Hylift Johnson offers special heavy-duty roller lifters (tappets) for non-MDS engines. These heavy-duty, premium lifters are designed to work with high spring pressure.

Remember that tappets *must* be installed before the heads go onto the block. The tappets cannot be serviced, removed, or replaced with the cylinder heads on.

Hydraulic roller tappets must be re-installed in the same tappet bore. The MDS lifter and the non-MDS lifter can't be switched in the guide. The alignment guide has two different-size guide holes.

Alignment Bar

When mechanical roller tappets were introduced into racing, typically a set of roller tappets had a guide bar that came with them designed to keep the roller running square to the lobe. In the Magnum engine in 1992 and newer, a figure eight–shaped dog-bone was used for two tappets. Gen III Hemis use an alignment bar, or guide, that accommodates four tappets at a time, so there are two per

side. These alignment bars are made of plastic. They bolt to a boss on the side of the tappet chamber. Jesel makes a metal replacement.

Camshaft and Spring Upgrades

Based on specific cam charts, engines are grouped by size, port size (the Eagle), and installed spring height. To switch an original 5.7

engine to longer valves that allow for higher lift cams is expensive. To switch the whole head assembly on an older 5.7 to the newer Eagle head to get the taller installed height and the bigger ports is even more expensive.

Performance Packages

A performance package is a list of all the basic parts that make up the engine: blocks, heads, cams, intake, exhaust, etc. It is important that all these parts work together to give the

Valve Package Upgrades				
Package	Valve Package Stage	Spring Lift (inch)*	Installed Height (inches)	Best Valvespring**
5.7 early and 6.1	1, 2, 3	.500	1.80	Stock and Comp Cams
	3, 4	.550	1.80	Comp Cams
5.7 Eagle†	3	.500	1.99	Stock and Comp Cams
	4	.550	1.99	Comp Cams
6.4/392	5	.590	2.00	Stock
	6, 7	.620	2.00	Stock and Comp Cams
	8	.675††	2.00	Comp Cams
426	6, 7	.600	2.00	Comp Cams or Stock
	8	.675††	2.00	Comp Cams
* As listed previously. ** Similar springs are available from Crane, PAC, PSI, and Manley. † 5.7 Eagle has a 1.99-inch installed height. Upgrade all 5.7 to Eagle heads; you can use the same cams as on the 6.4. †† Requires T&D high-ratio "race" rocker arms at this time.				

best performance in horsepower and torque. Because the cam is the last decision in the short-block and the cam's lift directly affects the cylinder head, this seemed like the best place to tie it all together: short-block, heads, intakes, exhaust, etc.

Following is a chart of the details of all eight available package options.

High-Performance and Race Cams

All production engines are hydraulic rollers, so there is no reason to select an old-school flat-tappet hydraulic or mechanical version. Mechanical roller cams are really an all-out race part and there are a few racers using them, but they tend to be custom and the tappets are not yet readily available.

To date, the hydraulic roller seems to work well in dual-purpose applications and even some all-out racing trials. This narrows your selection pool to hydraulic roller cams. The engine is still fairly new and buyer requirements are still evolving, but many profiles are available.

Comp Cams, Crane, and other aftermarket cams often offer similar features and benefits. Most cam manufacturers make .550- to .590-inch-lift cams; Comp Cams offers several hydraulic rollers in the .620-inch range. The higher lifts, such as .650 and .675 inch, are coming, and .750-inchers are likely to be offered soon. No mechanical versions are available yet, but they are currently being developed along with mechanical tappets. The testing and development hardware is performed on billet cams and most manufacturers offer billet cam service.

Background

The original 2003 5.7 Hemi was the base, everyday workhorse with a .472-inch valve lift and only 260 degrees of duration because of the hydraulic roller tappets. They allow pushing the design profile, so it's much shorter than the older flat-tappet designs.

A few years later, the Chrysler production engineers topped this (the 5.7 in 2009 and the 6.4 in 2011) when they introduced a long-valve package with a cam that has .571-inch valve lift and 286 degrees of duration. This may be the largest production cam ever. In the 1970s, 1980s, and 1990s, this would have been considered a race cam. The goal posts have been moved.

No Springs Cams

There are two basic groups of cams for Gen III Hemis: those designed for around .475-inch lift and a 1.800-inch installed height/production spring and those designed for the .570-inch lift and a 2.000-inch installed height. Some cams are designed for use with the stock valvesprings and the stock installed height (i.e., no spring changed required), which offer more street performance. Modern Muscle offers these special performance cams. Changing the cam and adding a new set of springs is much more expensive than just changing the cam.

Performance Packages			
Package Options	**Stage 1**	**Stage 2**	**Stage 3**	**Stage 4**
Displacement (cc)	347	347	347	347
Engine Years	2003–2008	2003–2008	All	All
Comp Ratio (:1)	10.2	10.2	10.2	10.2
Cam Lift (inch)	.480	.480	.500	.550
Duration at .050 (degrees)	195 to 210	200 to 215	205 to 220	210 to 225
Throttle Body	Stock	Ported	Ported	Big single*
Intake Manifold	Stock	Stock	Stock	Ported
Head	Stock	Ported	Eagle**	Ported Eagle
Intake Valve (inches)	2.00	2.05	2.05	2.05
Exhaust	Cat-back	Cat-back	Cat-back	Headers
Estimated HP/CI	1.07/1.12	1.12/1.22	1.20/1.35	1.30/1.44
Package Options	**Stage 5**	**Stage 6**	**Stage 7**	**Stage 8**
Displacement	392	392	392	392
Comp Ratio	10.7	10.7	10.7	10.7
Cam (Lift)	.550 to .590	.590 to .610	.600 to .620	.675†
Duration @ .050	220 to 230	220 to 230	225 to 235	235+†
Throttle Body	Ported	Big single*	Bigger††	4-Barrel
Intake Manifold	Stock	Ported	Ported	Hi-Rise
Head	Apache**	Ported	Ported	Big-Valve Ported
Intake Valve	2.14	2.14	2.14	2.20
Exhaust	Cat-Back	Headers	Headers	Headers
Estimated HP/CI	1.30 to 1.42	1.40 to 1.50	1.55 to 1.65	1.70 to 195

* Big single enlarges the throttle bore from 80 to 85/87 mm.
** This stage recommends a bracket valve job if the cam is changed.
† Biggest cam currently offered.
†† Bigger (single) moves to the 92-mm (Hellcat) throttle bore or 95- to 100-mm adapter.

High-Performance Hydraulic Roller Cams

With the Gen III Hemi there is no "Street Hydraulic" or "Mechanical" (both flat-tappets).

The hydraulic roller does a good job on the "street" side and the "drag race/competition" is not clearly defined at this writing.

Aftermarket manufacturers, such as Comp Cams and Crane, have mechanical roller profiles, but they haven't ground them on the Gen III cam blank yet (they have test samples only). Once buyers ask for them, the cam manufacturers will make them and then list them in their catalogs.

Stage	Company	Profile	Lift (inch; intake/ exhaust)	Advance Duration (degrees; intake/ exhaust)	Duration at .050 inch (degrees; intake/ exhaust)	CL (degrees)	Spring
			2003–2008 5.7* Cam Upgrade Packages				
Stock	Chrysler	Early 5.7	.472/.460	260/268	N/A	112	stock
AHR1	Comp Cams	112-525-11	.470/.464	246/258	194/206	116	26918
AHR2	Comp Cams	112-530-11	.477/.470	254/264	202/212	115	26918
AHR3	Comp Cams	112-535-11	.483/.477	262/270	210/218	114	26918

* Limiting factors are part and valve size, valve length, and valve spring.
5.7AHR1 = cam for performance package 1
AHR2 = cam for performance package 2
AHR3 = optional cam for package 2 or standard cam for package 3

The Eagle head (2009 and newer 5.7) uses a 1.99-inch installed spring height so it can use bigger cams than the earlier 5.7 engines.

More Lift

Most cam manufacturers make bigger cams that have more lift and more duration. Typically, big cams require big valvesprings, which can be an added complication and expense. Currently, most aftermarket cams have valve lifts in the .580- to .625-inch range. So far the high-lift winner is the .675-inch version used in the 426 Drag Pak. It works well and is based on the .620-inch valve-lift design; the production 1.66/1.68-ratio rocker arms are replaced with high-ratio (around 1.82) rockers designed by T&D.

Today

This .675-inch limit isn't going to stay long and may already be surpassed. As valve lift increases, any profile in this area is going to be an all-out race cam. For it to be successful, it must be easy to install, well developed, and very dependable. Developers often try several approaches and then select the best one. For street and dual-purpose engines, I use the .675-inch limit and show you other possibilities. It takes some valvetrain hardware (see Chapter 8 for details).

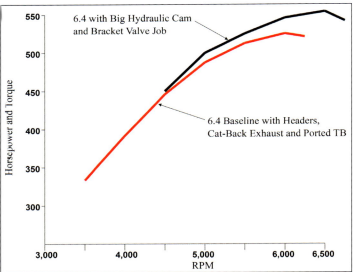

The best cam comparisons are those that make sense to the user. In that vein, I did not choose to make the cam change compared to a stock engine but to a baseline that has headers, an American Racing Headers (ARH) cat-back exhaust, and a Modern Muscle ported throttle body. These parts were run on a 6.4 engine for the baseline. The comparison was made to a bigger hydraulic cam at about 225 degrees at .050 inch and around .600-inch lift for a Comp Cam along with a bracket valve job. To install a cam, the head must be removed, so it seemed reasonable that a bracket valve job tune-up to the head would be likely. The result was 553 hp, or a gain of 44 hp. The gains would have been higher using a 5.7 Eagle.

| \multicolumn{8}{l}{2009 and Newer Apache 5.7 and 6.1* Cam Upgrade Packages} |
Stage	Company	Profile	Lift (inch; intake/ exhaust)	Advance Duration (degrees; intake/ exhaust)	Duration at .050 inch (degrees; intake/ exhaust)	CL (degrees)	Spring
Stock	Chrysler	New 5.7	258/288	NA	114	Stock	
BHR1	Comp Cams	112-535-11	.483/.477	262/270	210/218	114	26918
BHR2	Crane Cams	HR-208/297	.505/.505	268/274	208/214	116	99831
BHR3	Comp Cams	112-500-11	.522/.525	260/264	208/212	113	26918

* Advantages are bigger and longer valves for more installed height. The new 5.7 should have longer valves and a 1.99-inch installed height, but some sources indicate that the .472/.460 lift can be achieved in other ways.
BHR1 = cam for performance package 3
BHR2 = optional cam for package 3
BHR3 = cam for performance package 4

| \multicolumn{8}{l}{6.4 (with VVT)* Cam Upgrade Packages} |
Stage	Company	Profile	Lift (inch; intake/ exhaust)	Advance Duration (degrees; intake/ exhaust)	Duration at .050 inch (degrees; intake/ exhaust)	CL (degrees)	Spring
Stock	Chrysler	Standard 6.4	.571/.551	286/288	NA	114	Stock
CHR1	Crane Cams	HR-222/3236	.550/.550	280/286	222/228	114	99831
	Comp Cams	112-502-11	.547/.550	273/277	224/228	114	26918
CHR2	Comp Cams	201-424-17	.612/.605	266/279	218/226	114	26918
CHR3	Comp Cams	201-428-17	.619/.612	274/287	226/234	116	26918

* Advantages are longer valves, 2.00-inch installed height, and good valvesprings.
C = 6.4 and engines with VVT
CHR1 = cam for performance package 5
CHR2 = optional cam for performance package 5, standard cam for package 6
CHR3 = cam for performance packages 7 and 8

6.2 Supercharged Cams

Currently, no cams have been designed for a supercharged engine other than for the production Hellcat. Therefore, I recommend that you copy the Hellcat. Select CHR2 and CHR3 cams and adjust based on the Hellcat design.

On these two cams the exhaust lobe is too small in duration. On CHR2 the intake to exhaust duration difference is 13 degrees, about half of what is needed (based on the Hellcat). Hence, the exhaust lobe should be around 292 degrees (standard is 279).

The CHR2 cam's centerline is 114 degrees and that's fine, but the installed centerline should be 4 degrees early, or 109 degrees on the intake side.

The CHR3 is similar, a 300-degree exhaust lobe (standard is 287) ground on 114 centers and installed at 119 degrees intake. In either case, you can give up a little exhaust lobe lift to gain the desired duration.

CYLINDER HEADS

All current Gen III Hemi cylinder heads are cast from aluminum. Until a few months ago, that was going to be the end of my cylinder head discussion: just five production heads. During the last few months, two more heads have been released and will be available by the time this book goes to press. I am sure there are more casting versions and variations on the way.

To date, Chrysler has made five versions of the Gen III Hemi heads. Although the early 5.7 and the 6.1 look similar, the newer 5.7 Eagle, 6.4, and 6.2 have big intake ports and look similar to one another. All of these heads have dual spark plugs, so two round tubes stick up above each chamber. The original dual-plug approach was developed on the Gen II racing engine in NHRA Pro Stock. It was not very popular with racers because of its complexity, but they did not have the digital computers that are used with the latest hardware.

The two tall, round spark plug towers above each chamber are the easiest way to identify a Gen III Hemi head from other engines. The size of the intake ports can separate the early 5.7s from the big port 6.4s, but after that it is best to look at the casting numbers. Unfortunately, casting numbers can be difficult to read and 3s, 8s, 6s, and 9s tend to look the same.

Chrysler uses a 10-character system for casting numbers with the first eight being numbers and the last two being letters. However, do not assume that every letter combination has been used in production. For example, the 2005 5.7 engine has

The aluminum Gen III Hemi heads use a shallow, oval combustion chamber. This head has combustion chambers that are about the same size as used on the small-block wedge chambers, around 70 cc.

Cylinder Head Specifications					
	2002–2008 5.7	2009–today 5.7 Eagle	2005–2011 6.1	All 6.4	All 6.2
Valve size (inches; intake/exhaust)	2.00/1.555	2.05/1.555	2.075/1.585	2.138/1.654	2.138/1.625
Valveguide (inch)	5/16/.3125	5/16/.3125	5/16/.3125	5/16/.3125	5/16/.3125
Valve length (inches; intake/exhaust)	4.86/4.75	5.165/5.135	4.905/4.829	5.126/5.100	5.062/5.006
Spring Installed Height (inches; intake/exhaust)	1.81/1.81	1.99/1.99	1.87/1.772	2.051/2.016	2.051/2.016
Weight (pounds)*	30	30	30	27	27
* Weights are approximate for finish-machined aluminum heads.					
There may be more versions of these heads that are not listed because Chrysler seems to start the initial builds with one set of specifications that then evolve, revise, or update into another package. Some are limited production and some may be newer models because these engines are still in production.					

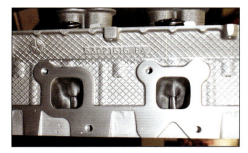

The exhaust port on the 5.7 engine is square. Also note the head's casting number between the tops of the two ports.

The D-shaped exhaust port on the 6.4 is wider than the 5.7's square port.

The 5.7 intake port is close to square with a dimple in the top edge, and the 5.7 uses a 2-inch valve. The 5.7 Eagle port (not shown) is larger.

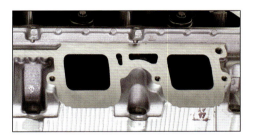

The 6.4 intake port is a very large square with a dimple, and it has large 2.14-inch intake valves.

Cylinder Head Casting Numbers		
Year	Engine	Casting Number
2003–2008	5.7 Hemi	53021300AJ/BA (right/left)
2009–2015	5.7 Hemi	53021616DD/DE (right/left)
2006–2011	6.1 Hemi	5037369AA
2011–2015	6.4 Hemi	5037369BD
2014–2015	6.2 Hellcat	5037369BD
This chart shows the 6.1 engine using the big-port 6.4 casting. This seems to be the SRT8 version in these model years.		
The Hellcat uses a different head assembly that has hollow intake valves and sodium-filled exhaust valves, but the head casting is the same.		

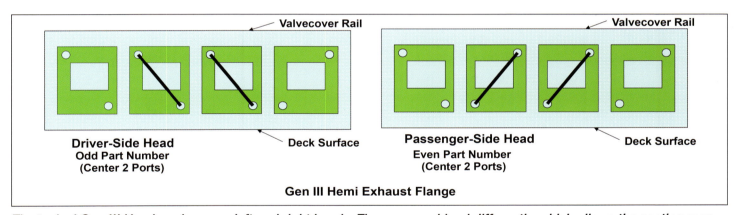

The typical Gen III Hemi engine uses left and right heads. They are machined differently, which allows the casting number to be the same even though there are left and right features. The main difference between the left and right head is that the two center ports on the exhaust flanges are machined differently: top left to lower right and top right to lower left. The four oil drainbacks are also machined differently.

a 53021300AJ casting number. The numbers represented by AA, AB, AC, AD, etc. may not have been used.

Aluminum Cylinder Heads

All Gen III Hemi heads are made of aluminum (including aftermarket versions), not cast iron. The twin-plug towers are the most obvious indication that the head is a Gen III Hemi. The size of the intake ports separates the big-port 6.4 versions from the 5.7s. Along with big ports comes bigger valves; several sizes are used and bigger valves could be added if desired. What is not so obvious but very important is that the big-port heads (6.4 and similar) have raised the rocker stand height, which is designed to use longer valves. This increases the valvespring installed height, which opens the door for much bigger cams.

This passenger-side head fits the 6.4. Note that the left port has the attaching bolts going top-left to bottom-right and the right port goes in the opposite direction. Only the center two indicate that it is a right-side head. The left-side head has the same bolt layout on the left and right ports but the center two go in the opposite direction, a left-hand head.

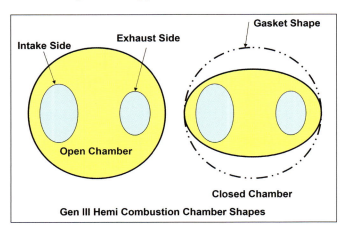

An open combustion chamber (left) is used mainly on the 5.7 engine; a closed chamber (right) is used mainly on the 6.4.

Combustion Chamber

Gen III Hemi engines use open and closed types of combustion chambers. The valve layout defines the hemispherical chamber with one valve on each side of the chamber. The open chamber has chamber area on both sides of the center oval defined by the valves or being open for 360 degrees around the top of the cylinder. The closed chamber fills in both sides so that the chamber shape is oval defined by the valve shape on each end. The 5.7 engine uses an open chamber and the 6.4 uses a closed chamber.

The size of the chamber is somewhat dictated by the valve angles used in the head. The Gen III Hemi uses an 18-degree intake valve and a 16.5-degree exhaust valve for a total included angle of 34.5 degrees. This is quite shallow. Gen II Hemi engines used a 59-degree included angle and had very large chambers. Gen III chambers are around 70. The 5.9 small-block chambers were around 70. The Gen III big chamber (open) is about 85 and the small (closed) chamber is around 65.

Dual Plug

All Gen III heads use the dual-plug layout, or two plugs per cylinder. If you draw a straight line between the centers of the intake and exhaust valves, the plugs are centered with one above the line and one below the line. There is no power difference between the two plug positions in the Gen III. The lower position in the Gen II made more power. Gen III spark plugs are 1-inch reach plugs rather than the 3/4-inch reach plugs used in small-blocks and Gen II Hemis. The high-performance plug for the Gen III engines is the Bosch SPHR5 or similar unit from Champion, NGK, or Brisk.

If you decide to switch the MPI system to a carburetor and use a distributor, similar to the Canadian circle-track Pinty's series, which requires a new front cover, you end up using a single-plug layout. Keep the same plug in the blank position (the one not connected to the

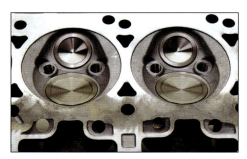

The 5.7's open chamber has two spark plugs: one on each side of the two-valve centerline.

The 6.4's closed chamber also has two plugs, but the sides of the chamber have been moved closer to the center to create a smaller, closed chamber.

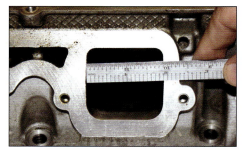

The 6.4 has very large intake ports that are 2 inches tall and 2.12 inches wide.

distributor). The power is the same with either position but the total spark advance increases from about 25 to about 35 degrees.

Valveseats and Guides

These heads use valveseat and valveguide inserts because the heads are made of aluminum. The valveguides (or valvestem diameters) are 5/16 or .312 inch, which is the same as on the Gen II Hemi. Using a valveseat insert gives you (and the manufacturer) a lot of flexibility in valve size without changing the casting. Material upgrades for these inserts offer performance advantages for racing applications.

Milling and Decking

Generally, you should not mill or deck the Gen III heads much; maybe .020- to .030-inch max. Do not mill the deck surface if you do not have to. Remember that alignment issues are related to the total amount milled off the head and the block, but volume is only related to head milling.

For example, if you have a closed-chamber Gen III head, you must mill .0068 inch per cc. Therefore, if your head is at about 67 and you want 65, you have to mill about .014 inch off the deck surface (2 x .0068 inch). On the other hand, if your machine shop took a .020-inch cut off the head's deck surface, you have lost 3.13 in chamber volume.

Port Size

One of the challenges with cylinder heads is the definition of a bigger port. Almost everyone wants bigger ports but what that means is often unclear. This is not a problem with Gen III Hemi heads because the ports on the 6.4 and its relatives are huge. The 5.7 and 6.4 intake ports are the same height (2 inches), but the 6.4 ports are much wider (2.12 versus 1.76). That's easy to see. The exhaust ports on the 6.4 are 1.70 wide and the 5.7's are 1.32 inches wide. Bigger ports use bigger valves and flow more air.

Restricting Oil to Heads

You should not restrict the oil to the cylinder heads, as was the common practice on Gen II Hemis. The factory engineers figured out the oil demand for the valvetrain and tappets and restricted the oil flow to this amount. This restriction is in the size of the hole in the head gasket where the oil passes. Further restriction causes problems.

Rocker Stands

The 6.4, 6.2, and the 5.7 Eagle (2009 and up) all have rocker stands subtly raised as part of the big-port package. The rocker shaft position is raised about .2 inch and widened about .140 inch. This height change allows for the use of longer valves and increased installed spring height.

Five rocker stand sets are cast into every head. One stand is on each end and the three center stands sit between each set of cylinders. A thick rib that runs vertically between the two plug towers ties the two rocker stands together. Also note the rocker shaft's oiling hole in the closer stand that intersects the attaching bolt hole and shaft saddle. The 6.4 head has a taller installed spring height because it uses longer valves. To retain the correct geometry, the rocker stands were raised. This 5.7 stand is just over 1/4 inch above the valvecover rail.

Cylinder Head Volume Ratio		
Amount to Be Milled (inch)*	Gen III Hemi Open Chamber (cc)**	Gen III Hemi Closed Chamber (cc)[†]
1-cc Ratio[††]	.0048 inch	.0068 inch
.010	2.13	1.56
.020	4.26	3.13
.030	6.38	4.69
.040	8.51	6.25
.050	10.64	7.81
.060	12.77	9.38

* The last 3 numbers, .040 to .060, are not recommended because of clearance issues.
** Most production 5.7 engines use an open chamber.
[†] All 6.4 and 6.2 Gen III engines have closed chambers, 6.1 may have issues.
[††] This ratio can be used to calculate the volume (cc) for any amount of milling.

This 6.4 head has the taller rocker stand, above the valvecover rail at about 1.0 inch.

When the stands were moved forward, they were also moved farther apart, but this is too difficult to see, so you must measure it using a steel scale.

There is no flat or machined surface at the bottom of the stand, but you can measure it more accurately with a steel scale than you can estimate it by eye. Note the dual plug towers in the center.

In addition, proper valve and rocker geometry is maintained. The rocker stands are cast into the cylinder head.

Spring Seats

Spring seats are machined into the aluminum head and a spring cup or shim should be used between the spring and the cylinder head. The 5.7 shim is flat and about .040 inch thick; the 6.4 version has an upside-down "T" shape. The spring seats are about 1.35 to 1.40 inch in diameter. This diameter dictates that the spring's outside diameter be less than this amount.

Exhaust Flange

The exhaust flange on Gen III heads is not the same on the left

If you place a head in this position (intake up, exhaust down), the oil drainbacks are next to the valvecover rail at the bottom. They are a little difficult to locate next to the two end head bolts, but the three in the center can be spotted. The lower row of head bolt holes is somewhat in line with the springs and below the five rocker stands. The drainback is next to the head bolt, just slightly to the left of the bolt face. There are only four in production heads and the left head is not like the right one; the front hole as installed is omitted.

and right cylinder heads. Although the castings are the same, there's a machining difference. The center two ports on each head are machined in mirror image. The top left is machined to the lower right with two attaching bolt holes; the top right is machined to the lower left with two attaching bolt holes. The end holes are the same on each head. Aftermarket heads have all four corners of the box machined around the two center ports so that they work on either side.

Drainbacks

Each head has four oil drainbacks, but the left and right heads are not the same. The drainbacks are located next to the exhaust flange. The front oil drainback on each head, as-installed, (the one all the way to the right on the right head and all the way to the left on the left head) is not machined on production heads. Aftermarket head manufacturers

machine all five drainback locations and provide a threaded plug for the undesired drainback.

Aftermarket Offerings

At the 2016 SEMA show, Edelbrock introduced a new aluminum

Edelbrock's all-new Gen III Hemi cylinder head was available in the spring of 2017. It is matched to the 6.4 with similar port sizes and valve sizes and the early flow numbers are similar. It uses a 2.165-inch intake valve.

The Edelbrock aluminum head has similar port sizes to the 6.4, but it's too new for any direct port comparisons.

The Edelbrock head has D-shaped exhaust ports that are similar to the 6.4's. Each exhaust port has all four corners around each port drilled for an attaching bolt (center ports), and this allows one head to be used on both sides. In addition, a fifth drainback hole is added with a plug to seal the one that you don't use (front as installed).

The CNC-ported head based on the production 6.4 casting (5037369 BD) uses stock-size valves and is the accepted head in the Canadian racing series that used a 362-inch version of the Gen III family for the 2017 season.

This TriTek CNC-ported head is similar to the basic 6.4 and based on a unique casting, which is also new. The CNC-ported flow numbers of more than 400 cfm with a 2.20-inch valve indicate good things to come.

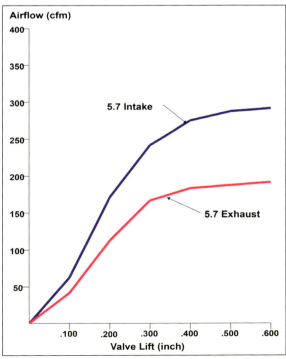

The early 5.7 head with its stock valve flows well. At about 285 cfm (intake), it is much better on the intake side than stock small-block (as-cast) heads such as the W2 or the Magnum R/T, which were thought to be the best stock small-block heads at the time. The W2 used a 2.02-inch intake so it was close. Good sources are Modern Cylinder Head, Modern Muscle, and Arrow Racing.

Gen III cylinder head, which became available in April 2017. It looks similar to the big-port 6.4 head and seems to flow similar numbers.

Racing Heads

Perhaps the most popular racing category for Gen III heads is the NHRA Stock and Super Stock classes for stock-based engines and 780-plus-hp Drag Pak engines based on the aluminum block. The Drag Pak engines use a CNC-ported head. The Canadian Pinty's circle-track and road racing series allows a 362-ci Gen III engine with a CNC-ported 2.14-intake and 1.65-exhaust valve 6.4 head.

In the X275 drag racing category (a 1/8- and 1/4-mile competition), the new aluminum head by TriTek holds the record (by Rob Goss) at 6.86 seconds and 205 mph. The racing TriTek head is CNC ported, but other versions are offered. The 2.165-inch intake flows 400 cfm (CNC-ported) and the 2.200-inch intake flows 410 cfm. Caution: These flow numbers may be with higher valve lifts of more than .600 inch.

Airflow

As racing developed, engine builders found that modifying the cylinder heads helped make the engine go faster. The problem was how to measure this gain. The answer was flow benches where only one cylinder head and one chamber on that head is tested at one time. This process was developed by the OEM factories and spread to aftermarket manufacturers and then to speed shop and engine builders.

In the early 1970s, only the three major OEMs had flow benches, but by the 1990s everyone had one. The trick to flow benches is to test all of

This 6.4 aluminum head is on a flow stand having its intake valve and port flowed. The micrometer at the center top is used to open the valve precisely.

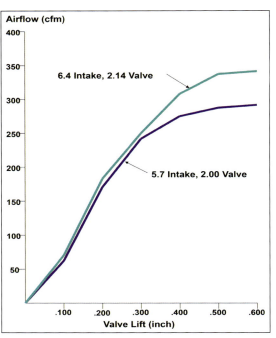

This flow curve compares the intake flow of the two major Gen III cylinder heads: the early 5.7 with a 2.00-inch valve and the 6.4 with its 2.14-inch valve. Both ports are stock. The 6.4 outflows the 5.7 by about 50 cfm at .600-inch lift. I probably should have flowed the 5.7 with a 2.14 valve, but it is too big for the port. A bigger valve (2.05 inches) helps, but it is still about 40 cfm short. Installing a bigger valve and CNC-porting the 5.7 head gets it closer, but that didn't seem fair. Modern Muscle, Arrow Racing, and Modern Cylinder Head provided the testing data.

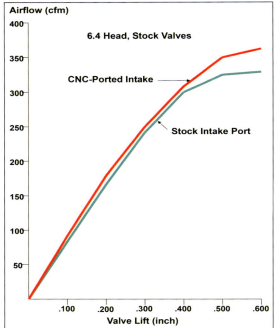

Ported cylinder heads provide lots of options. Many head specialists like to show flow numbers in .700-inch and higher lifts, but this didn't seem realistic because the current cams from Comp Cams and others are only getting into the low .600s (valve lift). Therefore, it seemed more useful to limit the lift/flow comparison to .600-inch lift. This curve compares the stock 6.4 intake port with a 2.14-inch valve with a CNC-ported head using the same valve size. The CNC-ported head flowed about 40 cfm better at the same lift. Good sources are Modern Muscle and Modern Cylinder Head, which provided the testing data.

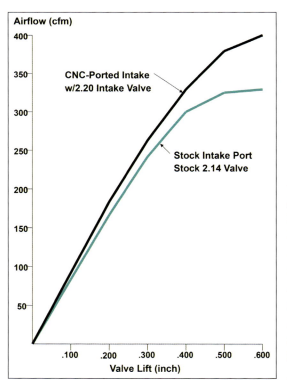

So far, the airflow champion of Gen III heads is the CNC-ported 6.4 casting with 2.20-inch intake valves. This package flows more than 400 cfm, which is 65 cfm better than the stock head. Modern Muscle, Arrow Racing, and Modern Cylinder Head provided the testing data.

the hardware on the same bench; do not end up racing flow benches.

In researching Gen III Hemi heads, airflow for the stock heads seems to be from a low in the 320s to a high of 340 cfm. Flow benches are accurate and would flow 340 every time, but switching benches introduces variations that the flow bench picks up to create higher or lower numbers. This situation is not unique to Gen III heads, but Gen III baselines are much less clearly defined than for older engines.

CNC Porting

Originally, special porting services techs ported heads by hand. Now, CNC equipment does almost all porting. The CNC machine should make all the ports the same dimensions, and that's very difficult to do by hand. With wedge heads, the CNC-machine has to make a left and right port; all Gen III Hemi ports can be the same. Typically, a CNC machine grinds and machines the valveseat at the same time as the port.

You can take one of two basic approaches to CNC porting. One is to design a port that fits the existing casting; the other is to design a casting for CNC porting. The five production heads are not designed for CNC porting so the port has to be designed to fit these existing castings. Edelbrock and TriTek heads may offer better CNC- porting capabilities. Remember that the CNC port is only as good as the prototype that was created by hand by a head-flow specialist.

With the 5.7 Gen III engine, Eagle heads (2009–2016) were so much larger than the original 5.7 heads (2003–2008) that you should consider these two engines separately. If you want

On this CNC-ported intake port in the 6.4 casting, you can see the airflow fin, or ridge, in the middle of the port.

Some head specialists shorten the cast-in valve guide and add or upgrade a longer valveguide insert (shown here sticking below the as-cast guide by about 1/4 inch). Shortening the valve's support below the port allows accelerated guide wear and more valve deflections, which are not good for street engines.

CNC-ported heads for the small-port (early) 5.7, use P5153345/346 (left and right), which are CNC-ported small-port heads with bigger valves (2.100 intake and 1.600 exhaust). If you want to use the big-port 5.7 Eagle, use P5160027; it flows 25 cfm (intake) better than the stock head.

The TriTek aluminum Gen III head focuses on CNC-ported versions rather than as-cast ports. These heads flow a lot of air, with 410 cfm for the big valve 2.200-inch version; the 2.165-inch version flows 400 cfm. TriTek also offers 2.08- and 2.125-inch versions, but they come

This CNC-ported exhaust port shows that the CNC-porting machine left some ridges, which is typical of this type of ported head. So far, the gains for ported intake ports are higher than they are for exhaust ports, but the 6.4 exhaust port flows well for a stock exhaust port.

Pushrod rod holes in the head are difficult to see. There are two per cylinder. Look at the dual-plug tower on the left; a pushrod hole is just above the left plug tower. The other one is to the right of the valvespring that's above the plug tower.

with a 1.655-inch exhaust valve, which flows 250 cfm.

Not all CNC-ported heads are created equal. For example, you could have a CNC-ported head that flows less than it did in stock form. You would like to assume that the porting gains will flow well, but it could have picked up 5 or 25 cfm. Which do you want?

The other side of the coin is how much do you want? The 25-cfm stock-valve CNC-ported head costs

Predicting Horsepower

SuperFlow (dynos and flow benches) has developed an equation for calculating an engine's potential horsepower output (race-equipped) based on the cylinder head's airflow numbers measured on a flow bench:

$$HPC = CP \times TF$$

Where:
HPC = potential horsepower per cylinder
CP = coefficient of power (for 28-inch test pressure flow bench, it is .26, per SuperFlow); this .26 number assumes gasoline for fuel
TF = test flow (peak) at 28-in pressure drop
Test flow is for one cylinder at max valve lift

For example, if the test head is a ported 6.4 aluminum Gen III head that flows 360 cfm at 28 inches of water, the formula calculates that the potential horsepower per cylinder is 93.6 (.26 x 360) or 749 for an 8-cylinder engine.

If you are using a .600-inch-lift cam and the head flows 360 cfm at .600 inch, 365 cfm at .650 inch, and 370 cfm at .700 inch, the 360-cfm number is used.

SuperFlow also developed a formula for predicting the peak power's RPM based on airflow numbers:

$$RPMPP = CPRPM \div (D \div 8) \times CFM$$

Where:
RPMPP = RPM at peak power
CPRPM = 1,196 for 28-inch test pressure
D = engine's total displacement
CFM = airflow in one port

Let's assume you have a 6.4, or 392-ci, engine and the ported 6.4 head with stock intake valve (2.14-inch) flows 330 cfm. According to the formula, the RPM at peak power is 8,054 (1,196 ÷ 49 x 330).

Keeping the same head, if you install a 4.00-inch crank for a 426-ci displacement, the RPM at peak power is 7,411 (1,196 ÷ 53.25 x 330.

Note that the longer stroke crank cost the engine about 600 rpm in peaking speed.

This formula does not work for supercharged or turbocharged engines. For example, a 6.2 head flows the same as a 6.4 head, approximately 330 cfm. Therefore, the potential horsepower per cylinder is 85.8 (.26 x 330) or 686 hp for an 8-cylinder engine. The Hellcat is already at 707 hp; no upgrades yet. So modified supercharged Gen III Hemis are at 720 hp and others are at more than 1,000 hp. Race versions depend on how much boost pressure is being used and the use of race gas. ■

---

a lot of money and the 40-cfm, big-valve package costs a lot of money not including valves. You don't need to pay big bucks for the 5-cfm package.

Obviously the intake valve (right) is larger than the exhaust valve, but the intake-valve underhead angle is shallow, almost flat (called a nailhead valve). The smaller exhaust valve (left) has a steeper angle (called a tulip valve).

Shrouding

The hemispherical combustion chamber allows valves to open into an open area on all sides except one: at the bottom next to the cylinder wall. For about 45 degrees, where the valve is next to the cylinder wall, the valve is shrouded as it lifts off the valveseat. The seat cutter may or may not hit this area to provide a relief. Enlarging this basic relief helps airflow. However, you must not get carried away because the seal ring on the head gasket must seal to the head surface in this area; any increased relief must stay inside of the gasket's seal ring.

Porting Advice

Do not try to make big-port 6.4 heads out of small-port 5.7 heads. There is not enough material. Be careful when installing big valves into small port heads even if they are ported. Do not forget the exhaust side when porting. It needs help too, especially the 5.7 versions.

Remember that porting is expensive. If you plan to use a stock lift cam, say .500 inch, you have no worry with the airflow at .600- or .700-inch lift. The valve will never get there. The .600- and .650-inch-lift

Both the intake and the exhaust valves on the 6.4 use three grooves. On the older wedge engines, the three-groove valve locks/keepers were used on exhaust valves only.

Cylinder Head Upgrade Packages				
Engine	Package	Best Head	Best Valves (intake/exhaust)	Installed Spring Height (inches)
5.7 early	Stage 1, 2	Bracket VJ*	2.00/2.05/1.56	1.80
5.7 early	Stage 2	Ported**	2.00/2.05/1.56	1.80
5.7 early	Stage 3	5.7 Eagle***	2.05/1.56	1.99†
5.7 Eagle	Stage 3	Bracket VJ*	2.05/1.56	1.99
5.7 Eagle	Stage 4	Ported	2.05/1.56	1.99
6.4/392(5)	Stage 5††	6.4 Apache	2.14/1.65†††	2.00
6.4/392	Stage 5	Bracket VJ*	2.14/1.65†††	2.00
6.4/392	Stage 6/7	Ported	2.14/1.65†††	2.00
6.4/392	Stage 8	Ported BV	2.20/1.65	2.00

* A bracket valve job is recommended if the cam is changed.
** Porting an early 5.7 head should be less expensive than swapping on a set of Eagle heads.
*** Optional upgrade readily available new or used; standard head is okay.
† The 1.99-inch installed height is the main reason to switch applications.
†† The Stage 3 package is really any 6.1 engine; they can use the Apache head.
††† The Hellcat has hollow intake valves and sodium-filled exhaust valves; not required for the naturally aspirated engine.

Ported heads, which means CNC-ported heads, are recommended in nearly 50 percent of packages. Sources for CNC-ported heads are Modern Muscle, Arrow Racing Engines, Advance Motion CNC/Darren Shumway, Modern Cylinder Head, and Indy Cylinder Heads.

The bracket valve job is shown as recommended with a cam change because to change the cam, you must remove the cylinder head. Because it is removed anyway, the bracket valve job is a good investment even if you do not plan to port the heads.

flow numbers become important if you use .650-inch-lift cams. (Refer to page 86.)

The Engine "Team"

Once the head has been selected, prepped, and ported, the finished head must work with the rest of the engine to form a team. The head's combustion chamber must be cc'd so that you can accurately calculate the engine's compression ratio. To do that accurately, you must calculate the engine's displacement.

CC-ing

The first step in measuring your engine's compression ratio is to CC the head's combustion chamber. Initially, you only need to measure one chamber. I recommend one of the two center chambers in the head. To CC the combustion chamber in the head, you need a 100-cc burette, a flat and clear 1/4-inch-thick plexiglass plate with a small hole in one corner, a cc-ing fluid (such as parts cleaning solvent or rubbing alcohol with red food coloring), and a light grease (such as petroleum jelly).

Next, install two spark plugs, seal the valveseats on the valves with light grease, and install the valves into the selected chamber. Seal the plate with a bead of light grease so that it seals to the top of the chamber. Install the plate over the chamber; the fill hole should be at the high spot. Fill the burette with cc-ing fluid and zero it. Fill the chamber and record the volume.

Compression Ratio

You want to know the basic compression ratio as soon as possible so you can order pistons, mill decks, mill piston tops, or order new pistons with the compression ratio your application and fuel use. See Chapter 4 for more details about calculating the engine's compression ratio.

You can't calculate compression ratio until you CC the head. The fuel, or gas octane, dictates what you should have for a compression ratio.

I placed this discussion here because the short-block is basically together (choice-wise) and the head is the final decision. I look at this as the last chance to fix it. Once you bolt the head onto the short block, it becomes difficult and expensive to fix any problem with compression ratio, either too much or too little.

Octane Influence

Before you cc the chamber, you may not be able to accurately calculate the engine's actual compression ratio. Gen III Hemis use electronic MPI, which enables the base, or stock, 5.7 engine to run a CR of 10.2:1 (almost one full point higher than earlier) on pump gas. Add to this the newer 6.4 engine and its 10.7:1 compression ratio. The most amazing of all is the Hellcat with 9.5:1 CR and supercharged.

Displacement Calculation

The first part of the compression ratio equation or VBDC, the volume above the piston at BDC, is basically the displacement of one cylinder plus the VTDC. This makes it important to calculate your exact cylinder displacement. You can use the following formula:

$$D = N \times .7854 \times B \times B \times S$$

Where:
D = displacement
N = number of valves
B = cylinder bore
S = cylinder stroke

For example, if your engine has a 4.09-inch bore and a 3.72-inch stroke, the displacement is 390.99 ci (8 × .7854 × 4.09 × 4.09 × 3.72).
For use in your compression ratio calculation, you only want one cylinder's displacement (sometimes called swept volume); in this example, 48.87 ci (390.99 ÷ 8). To use this number in the compression ratio formula, you need to convert the volume to milliliters 800.89 (48.87 × 16.387).

I recommend using a thick head gasket or dropping the piston to achieve a 9.5:1 CR (for supercharged engines). However, the ability to use these high CRs on pump gas has to be credited to the dual knock sensors, the combustion chamber design by Chrysler engineers, the intercoolers used on the supercharged engines, and the basic MPI engine map used in these engines. Do not disable the knock sensors on these engines.

The fuel is the key to the amount of compression ratio that you can use. Factory engineers have done a great job of getting the most out of pump gas. If your engine project is slated to be a street or dual-purpose project, you want to use gas that is readily available and available for a reasonable price. Typically, this means pump gas. Some general guidelines are as follows:

- Pump Premium (about 92 octane): about 10.7:1 CR with aluminum heads and MPI; about

9.5:1 maximum with a street supercharger and aluminum heads and MPI
- Race Gas (over 100 octane): purchased in 55-gallon drums or at the racetrack; for race cars (expensive) and race engines only, 11:1 CR and up or over 10:1 with a street supercharger.

If you are shooting for a 10:1 CR and you find that the engine has an 11:1 CR, you must lower the ratio by one point. When you are building, a new set of pistons is the best solution. If it is already built, you should try a thick head gasket or shim.

Seats and Guides

To repair an aluminum head's valveguide or valveseat, the damaged guide or seat must be removed and replaced by an oversize guide or seat. The outside diameter of the insert is oversize also. Oversize parts are available from the aftermarket.

Valves

The pistons and piston rings seal the bottom of the combustion chamber. The valve's job is to seal the top of the chamber and also let in the induction flow and let out the exhaust flow. The big change in Gen III Hemi valves is the significant jump in length from the early 5.7 to the newer 6.4 and its relatives.

Length

The 5.7 Gen III Hemi's intake valve has a length of 4.86 inches and the exhaust length is slightly shorter at 4.75 inches. The 5.7 Eagle (2009 and newer) and the 6.4 engines use valves that are about .300 inch longer at 5.165 intake and 5.135-inch exhaust; the 6.4 is about .030 inch shorter than the 5.7 Eagle. This extra length allows the valvespring installed height to increase and this allows for bigger springs and more valve lift. This same technique has been used on racing small-blocks since the 1970s.

Valve Specifications					
	2003–2008 5.7	2009+ 5.7 Eagle	2005–2010 6.1	2011+ 6.4	2014+ 6.2
Valve size (inches; intake/exhaust)	2.00/1.555	2.05/1.555	2.075/1.585	2.138/1.654	1.92/1.625
Valveguide (inch)	.3125	.3125	.3125	.3125	.3125
Valve length (inches; intake/exhaust)	4.86/4.75	5.165/5.135	4.905/4.829	5.126/5.100	5.062/5.006
Spring Installed Height (inches; intake/exhaust)	1.81/1.81	1.99/1.99	1.87/1.772	2.051/2.016	2.051/2.016

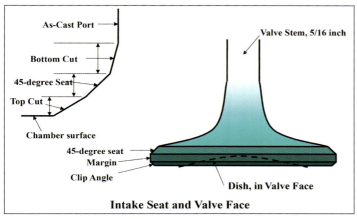

Intake Seat and Valve Face

The intake valveseat and the seat in the head itself should both be ground carefully. The Hemi has always liked the multi-angle (or full radius) approach angles for the valve job. Do the exhaust valve the same way with smaller cutters.

If the head is off the engine, I recommend that for the first round of modifications you do a bracket valve job. It involves backcutting the valves, so the angle is about half the seat angle. The problem with this approach is that the backcut makes a 45-degree seat narrower. It flows better but is not as durable on the street.

Stem Diameter

All Gen III Hemis use 5/16-inch valvestems. This is the same size that was used in Gen II 426 engines. Most small- and big-blocks use 3/8-inch; 11/32 is popular in the aftermarket.

Head Diameter

The original 5.7 Hemi engine used 2.00-inch intake valves and 1.555-inch exhaust valves. These were slightly smaller than the 2.02 intake and 1.60 exhaust valves used in the high-performance 340 in 1968–1971. The 5.7 Eagle (2009 and newer) has a bigger intake valve (2.05 inches) to go with the bigger ports in the head.

The 6.1 engine used .025/.030-inch bigger valves than the Eagle. The big valves are used in 6.4/392-inch engines, which receive almost .100-inch bigger intake (2.14 inches) and exhaust valves (1.65 inches) to go with the much bigger ports.

Oversized valves are available for the intake with 2.165 and 2.200 being most popular. More sizes will be available as demand increases and machine shops find the sizes that work best with the basic port sizes.

Grooves

The valve locks, or keepers, require grooves to be cut in the valvestem. Almost all Gen III Hemi engines

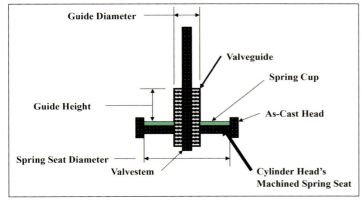

The production guide height above the machined spring seat is suitable for most current applications. However, as valve lift goes over .600 inch, the height of the guide in the head may have to be shortened to allow for any extra lift. All aluminum Gen III heads use a shim or spring cup between the spring and the cylinder head. The 5.7 uses a flat shim; the 6.4 uses an inverted T-shaped spring cup.

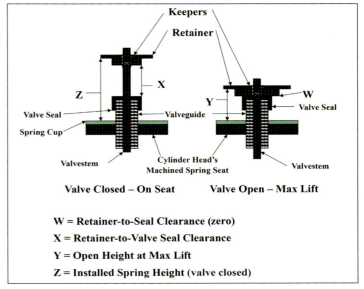

Valve Closed – On Seat   **Valve Open – Max Lift**

W = Retainer-to-Seal Clearance (zero)
X = Retainer-to-Valve Seal Clearance
Y = Open Height at Max Lift
Z = Installed Spring Height (valve closed)

As valve lift increases, you must pay close attention to the various seal and retainer clearances based on the guide height.

Although they are not all exactly the same, production Gen III engines use a black plastic valvecover that's fairly flat or shallow.

Arrow Racing makes a cast-aluminum valvecover that replaces the plastic cover.

An Arrow Racing cover (right) is about 1 inch taller than a production plastic cover (left).

Modern Muscle offers this fabricated aluminum cover.

Moroso offers this billet aluminum valvecover for Gen III engines.

use a single groove in the valve. The exception is the 6.4 engines, which use three grooves on both intake and exhaust.

Material

Both production intake and exhaust valves are made from steel. In general the intake and exhaust are not made from the same steel alloy. The exhaust valve sees more heat so it receives a special alloy. Most aftermarket performance valves are made of stainless steel. The 6.2 Hellcat uses special high-strength materials in its valves for use with a supercharger.

For max-performance applications, the aftermarket offers titanium valves, which are considered race-only parts. They are strong and lighter than stainless valves, but they are quite expensive. Because valve heads are larger and valves are longer, they become heavier and titanium is one way of lightening the valve without giving up diameter or length.

Valvecovers

Production valvecovers are black plastic. Aftermarket covers are cast-aluminum and billet aluminum. They fit on either head.

Plastic

Most Gen III Hemis use a black plastic valvecover. The 6.4 engines use a special plastic cover, but it looks the same on the outside. The differences are that some ribs are moved to the inside of the cover.

Cast Aluminum

Arrow Racing Engines offers a cast-aluminum cover, which is taller, and comes with longer plug boots and special tower seals to make it a complete package. The taller cover gives more room for the big valvesprings, longer valves, and race rocker arms.

Billet

Currently Moroso and Modern Muscle offer billet aluminum valvecovers. Many options are available with billet covers.

Head Gaskets

Gen III Hemis use the same gasket on both sides and should be installed with the "UP" designation on top, next to the head, not the block deck.

The 5.7 engines use a .027-inch-thick gasket with a small bore size; the 6.1 and 6.4 engines use a thicker .040-inch gasket. The key with all these engines (including the big-bore 426) is that each larger bore size receives its own head gasket with the matching bigger bore in the gasket.

Cometic Gasket offers big-bore gaskets for the 426 (over-bore from the standard 4.125-inch).

All of these head gaskets use multi-layer-steel (MLS) construction. Although each manufacturer varies the details, the MLS gasket design is based on a sandwich-style assembly with a steel layer in the middle and a sealing layer on top and bottom. MLS gasket designs work well for performance applications.

Cometic offers MLS Hemi gaskets in custom sizes (thicknesses) up to .140 inch.

Head Bolts

Gen III Hemis use a unique 6-bolt head-attaching pattern. There are 4 long bolts around each chamber and another 2 smaller bolts per cylinder at the top of the tappet chamber. The big bolts are M12s; 10 per head. The smaller bolts are M8s; 5 per head. Each head is aligned by two, large-diameter, hollow dowels around the numbers-8 and -10 head bolts (see Chapter 2 for more details).

Bolts or Studs

In racing applications, where it is common to remove the cylinder head frequently, replacing the head bolts with studs is common. Studs can also offer advantages for torque loading. ARP has bolt and stud kits for Gen III engines.

Caution: On a street engine with master cylinders and other items in the engine compartment that are near the engine, studs require the head to be lifted straight up over the studs and this may not be possible in the car.

The production engine uses 10 long bolts (right) plus 5 smaller bolts (along the top) per head.

ARP offers head studs to replace the production bolts. The studs, nuts, and washers come wrapped in heavy plastic or bags to keep them from banging each other and to keep other bolts and nuts from hitting them.

Head Gasket Specifications			
Engine	Stock Bore (inches)	Gasket Bore (inches)	Thickness (inch)
5.7	3.917	3.95 and 4.10	.027
6.1	4.05	4.10 and 4.125	.040
6.4/392	4.09	4.12 and 4.15	.040
426	4.125	4.125, 4.185, and 4.25	.040

The chart is based on the Cometic catalog.

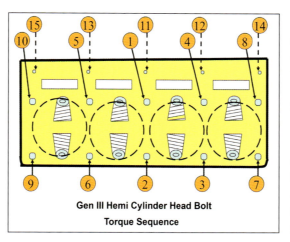

Gen III Hemi Cylinder Head Bolt
Torque Sequence

The large M12 bolts are marked 1 through 10 and should be torqued first in a three-step sequence. Step 1: Torque the M12 bolts to 25 ft-lbs. Then torque the M8 bolts (numbers 11 through 15) to 15 ft-lbs. Step 2: Torque the M12 bolts to 40 ft-lbs and then torque M8 bolts to 15 ft-lbs. Step 3: Turn the M12 bolts 90 degrees in sequence and torque the M8 bolts to 25 ft-lbs.

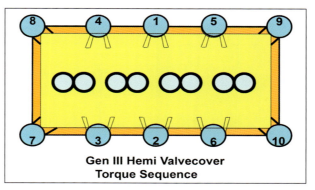

Gen III Hemi Valvecover
Torque Sequence

Torque the valvecover screws to 70 in-lbs in sequence.

VALVETRAIN

Chrysler engineers did an excellent job designing a Gen III valvetrain that, in stock form, can support up to 600 hp. With some modification and aftermarket parts, it can support much more for a max-performance build. However, with horsepower and torque performance targets in mind, valvetrain upgrades need to be considered when upgrading to a higher lift with different timing specs.

Ultimately, the valvetrain transfers the cam motion to the valves, and thus, the valvetrain is all of the parts between the lifters/tappets and the valves in the heads. Valves are discussed in Chapter 7 because the valveseats in the head are only half the story. The valveseat is the other half, so it makes sense to discuss them together.

Lifters are covered with the camshaft in Chapter 6. With Gen III Hemi engines the tappets must be installed before the head so that they are also tied closely together.

The valvetrain includes the pushrods, rocker arms, rocker shafts, valvesprings, retainers, keepers, and valve seals. Rocker stands could be a concern, but they are cast into the heads.

Hydraulic and mechanical are the two types of valvetrain. The camshaft and its tappet define a particular type. So far, there is no mechanical valvetrain for the Gen III engines. Without a mechanical system, you could still consider the two types of valvetrain: roller and flat-tappet.

All Gen III engines use a hydraulic roller system, so backing up to a flat-tappet system doesn't seem realistic. With .571 inch of lift in production, the basic hardware has to be on the leading edge. A mechanical system starts with adjustable rockers and the T&D Race system is adjustable.

The MDS system and the VVT system do not appear to cause any problems for the valvetrain. Although the cam and tappets change with both, the rest of the valvetrain doesn't change.

Gen III engines oil the valvetrain up through drilled passages in the head and block. They oil the internal parts of the hydraulic tappet down the pushrod. The valvetrain oiling restriction is built into the head gasket. The oil moves into the rocker shaft, up and down to each rocker arm, out to the cup in the rocker arm, and into the hollow pushrod through the hole in the pushrod's end cup.

The valvetrain area is a rapidly developing group of products. Chrysler engineers did an excellent

Most of the engine's valve gear installs on top of the cylinder head. Once the rocker and springs, shafts and hardware are installed, it's pretty crowded.

Although the intake and exhaust Gen III rockers may look similar, the intake valve tip end is clearly marked with an "I." Also the rocker slides down the shaft to uncover the oil feed holes to the rocker, which is how oil travels to the valve tips, pushrod, and tappet.

job of designing and developing the standard valvetrain package so it can go racing.

For example, an output of 700 or 800 hp for the small-block was a full race engine a few years ago (and a very competitive one), but the Gen III Hellcat comes with more than 700 from the factory, fully-certified and emissions legal. If you swap a pulley and tweak the ECM (sometimes called re-programming), it could easily be at 800 hp. It could be 1,000 hp soon.

Pushrods

The pushrods are 5/16-inch in diameter and have a pivot on each end so they can be installed in either direction. Oil travels down the push-rod to make the hydraulic tappets work properly. Intake pushrods are about 1.25 inches shorter than the exhausts; the long ones go to the exhaust. The 426 aluminum-block package (Drag Pak and others) use a stiff 3/8-inch pushrod that is double tapered. As cam lifts and spring pressures increase, stiffer pushrods are recommended.

Rocker Arms

The intake and exhaust rocker arms are unique. The four intake rocker arms mount on a rocker shaft while the four exhaust rockers mount to a separate shaft.

Valvetrain Specifications			
	5.7/6.1/6.4/392	6.2 Hellcat	426 Aluminum
Rocker Shaft OD (inch)	.865	.865	.865
Rocker ratio (inches; intake/exhaust)	1.60/1.66	1.60/1.66	1.80/1.85*
Tappet diameter (inch)	.8420 to .8427	.8420 to .8427	.8420 to .8427
Pushrod, hydraulic OD (inch)	.309 to .315	.309 to .315	3/8 HD*
Pushrod length (inches)	6.60** intake	6.60** intake	6.775 intake
	7.85** exhaust	7.85** exhaust	8.075 exhaust

* High-ratio rockers and 3/8-inch heavy-duty standard on Drag Pak SS package.
** There seem to be more than these two lengths: raised shafts, SRT8, etc.

The Gen III Hemi engine uses two lengths of pushrod. The long ones go to the exhaust side, and the difference is easily visible.

High-performance engines, such as the 426 Drag Pak or crate engine, use a double-tapered 3/8-inch pushrod (production is 5/16 inch) made by Trend. They are also slightly longer than the production version.

The two rocker arms are quite short, which makes them stiff but still light. The intake and exhaust versions look alike, but the intake rocker has an "I" on the top of the valve tip arm. The rockers, made of investment-cast steel, are strong and stiff.

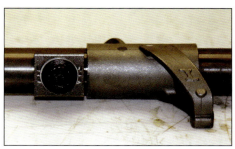

Production rocker arms fit against the shaft retainer and do not have spacers. However, the high-performance T&D set uses spacers.

This T&D race rocker system has valve-lash adjusters and can be used with mechanical tappets and cam. A roller tip is used, which comes in several ratios. They need added clearance to the plug tower.

T&D offers this street system with and without valve adjusters. Both use the stock ratios of 1.60/1.66 (intake/exhaust). The street and race designations are a way to identify one set or style from the other.

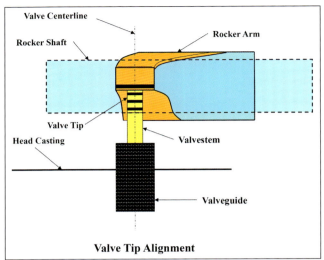

Valve Tip Alignment

The production rockers and valves align precisely. However, if you start swapping and modifying parts, you need to verify that the rocker tip and the valve tip are properly aligned.

High-Ratio Rocker Clearance

So far, only the T&D race system has been offered in a high-ratio rocker, which is 1.80 to 1.85 compared to the stock 1.60 to 1.65. It is used on the 426 Drag Pak engine to gain about .675-inch valve lift and is used with a bigger valve-spring from Comp Cams. These high-ratio rockers require that some clearance be ground into the spark plug tower.

	5.7, 6.1, 6.4	6.2 and 354	426	426 Drag Pak
Rocker Arm Ratio (:1)	1.60 to 1.66	1.60 to 1.66	1.60 to 1.66	1.80 to 1.85
Rocker Shaft OD (inch)	.865	.865	.865	.865

Rocker Ratio

The typical stock Gen III rocker ratio is 1.6 to 1.66. A replacement rocker package from T&D is available in the same ratio. A race rocker system made by T&D comes in ratios up to 1.80 to 1.85. Jesel also makes a race rocker system for Gen III Hemis, which uses mini–rocker shafts. Jesel has introduced a replacement rocker system that uses stock rocker shafts.

Rocker Shafts

Each Gen III Hemi cylinder head has two rocker shafts. One is on the intake side (top) and the other one is on the exhaust side (bottom). It is a long, thick-wall, hollow tube with holes drilled in it. Production

Production rocker shafts and rocker arms need to be stiffened up if stronger valvesprings are installed. Use tie bars for performance applications where bigger cams with more valvesprings loads are being used; I recommend that you use a tie bar across the tops of the rocker stands. There are two versions: one for the early 5.7 and one for the raised-shaft 6.4 and family. They are available from Mopar Performance. Tie bars are not required with the T&D system.

shafts are ground round. Each shaft is attached to the head by five screws into cast-in rocker stands. The shaft OD is .865 inch. The valvetrain is oiled from the head up and down the shaft to the rockers and out to the valve tips and pushrods.

Production rockers do not have any valve-lash adjustment; the aftermarket versions do have adjustment; replacement versions may offer options. Remember that the 6.4 and its tall installed spring height heads have the rocker shafts raised about

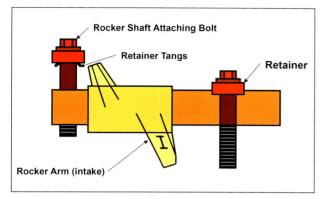

If the engine has been assembled, I recommend that the rocker shaft attaching bolts not be removed from the shaft. Two tangs on the bottom side help hold the shaft and bolt in place. Tangs might be damaged or the special washer lost if the bolt is removed.

The advantage of the beehive spring is that the retainer is much smaller than a standard retainer and therefore much lighter. The beehive spring also does not need a dampener inside the main spring because the tapered design of the spring dampens itself. The spring cup or shim (6.4 style) is at the lower right and left.

.200 inches and spread apart by about .140 inch. If high-ratio rocker arms (race versions) are being used, the spark plug towers must have some clearance added.

Before you install the rocker shaft assemblies onto the head, rotate the crank about 45 degrees from TDC so that all pistons are halfway down the cylinders. This helps avoid piston-to-valve contact during the assembly process.

Rocker Stands

Rocker stands are cast into the cylinder head. Two versions are available: standard 5.7 (early) and 6.4 and family. The 6.4 is about .200 inch taller and about .140 inch farther apart. A thick rib is cast into the head that connects the rocker stands. This feature provides more support for high-performance valvetrains.

Valvesprings

All production Gen III Hemis use beehive valvesprings. These are tapered so that the top is narrower than the bottom. Several other styles of springs, such as single springs, dual springs, and triple springs, have been used on other engines or in racing. The most common production spring used to be the single, but now

it is the beehive. In the early and mid-1960s dual springs were used in some high-performance engines, including the early Gen II 426 Hemi, but have been rare in production since about 1970. Triple springs are a race-only part.

Beehive springs started appearing in production engines in the early 1990s; they were used on the Magnum (small-block) engines in the early 1990s. At the time, only the beehive spring fit the head, which had almost no options, so Mopar Performance designed a straight spring to replace it. Today there are many beehive options.

A valvespring is designed to work at a specific installed height and has specific outside and inside diameters.

Almost all valvesprings were

straight (same size top and bottom) until the beehive came along in the early 1990s (in production). For the first few years, wire-bending technology was too new and optional springs for this configuration were almost non-existent. To solve this problem, Mopar introduced straight, single springs that fit the Magnum machined-spring seats in the heads. Since then, Crane, Comp Cams, PAC and Manley have introduced several performance options for beehive-style springs.

Production Valvespring Specifications				
Engine	5.7	5.7	6.1	6.4 and 6.2
Years	2003–2008	2009 and up	2005–2010	2011–2016
Installed spring height (inches)	1.771	1.99	1.87/1.77	2.051
Seat load (pounds)	97.8	116.9	99.0	114.7
Open height (inches)	1.283	1.47	1.30/1.22	1.48
Open load (pounds)	242.0	331.6	325.5	337.2
Rate (pounds)	295	413	397	389
OD, bottom (inches)	1.35	1.35	1.35	1.35
Although the seat loads are similar to earlier V-8 engines, the open loads over 320 pounds are impressive. The high-performance 340 and 440 engines in the muscle car era had open loads in the 230- to 250- and 260- to 280-pound area, respectively. It's another advantage of roller cams.				

High-Performance Beehive Valvespring Specifications				
Company	Chrysler	Comp Cams	Comp Cams	Comp Cams
Part Number	6.4 Stock	26918	26120	26095
Style	Beehive	Beehive	Beehive	Beehive
Installed Spring Height (inches)	2.051	1.800	1.880	2.000
Seat load (pounds)	114.7	125.0	155	150
Open height (inches)	1.48	1.15	1.28	1.25
Open load (pounds)	337.2	367.0	377.0	375.0
Rate (pounds)	389	372	370	300
OD bottom (inches)	1.350	1.310	1.445	1.589

I have deliberately recommended Beehive springs. There are enough options today to cover most requirements. Dual and triple springs are readily available, but that is old technology and possibly more expensive because the retainer must also be replaced. Another problem is that dual springs commonly have a 1.5 OD, which is too large for the machined seat in the head and expensive to fix.

There are a lot of advantages to using a beehive spring, but perhaps the most obvious is that the valvespring retainer is much smaller and therefore lighter. If you are going to use more valve lift, the first thought is the production 6.4 spring; stock lift is .571. Remember that the 5.7 (early version) uses a .480 lift cam and a 1.80 installed height.

Both Crane and Comp Cams offer an upgrade in valvesprings for their typical performance cams, about .590- to .615-inch lift.

The .675-Lift Package

So what is your next option? So far, cams listed in the catalogs do not offer valve lifts as high as .675-inch, based on a 1.6 rocker ratio. This higher lift package was accomplished by changing the rocker ratio from about 1.65 to 1.82. The original cam had an approximate .615-inch valve lift with stock-ratio rockers.

A Comp Cams spring (26120) installs at 1.88 inches. For .675 lift, the installed height can be increased to 1.98 inches because the 6.4's stock installed height is about 2.00 inches. This drops the seat load to about 118 pounds, which is slightly higher than the stock 6.4's 114.7.

If a cam is more aggressive, the 115- to 120-pound seat load will not be enough and you will need a new spring. The Comp Cams spring (26095) has the lift capacity and 150 pounds on the seat but has a 1.589 OD, too big for the machined spring seat in the head.

Soon, cams may be at .700- to .720-inch lift. Based on the specifications, the Comp Cams spring (26120) installed at 1.98 with 118 pounds on the seat could lift up to .700 inch. Other springs are in the same category; for example, springs from PAC and PSI.

So what if you want to go to a .750 lift? Today, you have one solution: the Manley beehive spring (221439).

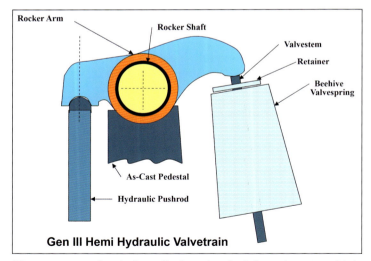

Gen III Hemi Hydraulic Valvetrain

There are two basic installed spring heights: the 5.7, which is about 1.80 inches, and the 6.4, which is about 2.05 inches. The current beehive performance springs can have up to .625-inch lift pretty easily and even .675 inch with careful selection (requires 1.8 ratio rockers at this time).

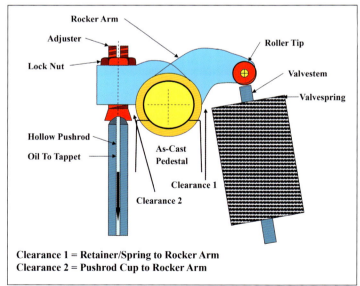

Clearance 1 = Retainer/Spring to Rocker Arm
Clearance 2 = Pushrod Cup to Rocker Arm

Mechanical rockers and straight springs must be checked for clearance (such as under the rocker arm) to the spring and pushrod.

The upscale step in rockers from the standard hydraulic production rockers is one of the two T&D systems; this is the race system.

Jesel offers a race rocker arm package. It is also a roller tip and roller pivot but is more complicated to install.

It is rated at 150 pounds at 1.80 inches with an OD of 1.31 inches. If this installed height is increased to 1.90 inches (using a shim on the stock 2.00 height) the seat load is 114.6, which is the same as on the stock spring. The rated open height is 1.15 inches so that means it can handle a .750-inch valve lift (1.90 - 1.15 = .750). New beehive springs may be introduced soon that have larger seat loads and higher lift capabilities.

A Racer's Answer

The solution for racers wanting to achieve higher valve lift would be to change to a standard straight spring. In that case, if they are looking mainly for larger seat loads and higher lift capacity, such a straight spring would be a dual.

The Manley dual spring (221446SF) has an OD of 1.400 inches. The Gen III spring seat is around 1.41 inches. At an installed height of 2.050 inches it has 240 pounds of load and at 1.250 open height it has 700 pounds of load, or a .800-inch lift. For the dual-purpose engine, this approach requires new springs and new retainers, an added expense.

Small-diameter dual springs are available from Manley, Comp Cams, Crane, PAC, and others.

Performance Springs

The cam lobe and the rocker arm ratio dictate the valve lift and the valve lift dictates the valvespring that is required. Production springs are designed to work at production valve lifts around .480- (early 5.7) to .580-inch valve lift (6.4).

Performance springs are designed for high-valve lifts in the .500- to .620-inch area, which could also be called street use. Gen III engines planned for the dual-purpose category probably have valve lifts in the .550- to .675-inch range.

Jesel introduced an all-new system, which he called a "replacement" system; it costs less than a standard Jesel Race setup. In addition, it's easier to install and less machining is required.

Valve lifts over .675-inch are considered race levels today, but not for long.

Typically a cam is designed to make more power (bigger lobes) and a bigger cam dictates a new valvespring. Modern Muscle has taken a somewhat unique approach and designed bigger cams to make more power. These cams are compatible with production springs.

Spring Loads

Each valvespring has two spring loads: closed, or seated, and open. For early 5.7 Gen III engines, the average on-the-seat, or closed load height is 1.80 inches. The seat load at this height is slightly less than 100 pounds. Other Gen III engines, especially the 6.4, use a 2.00-inch closed height or installed spring height. The load on the 6.4 at this closed height is slightly less than 115 pounds. You measure the open load at the installed height minus the max valve lift.

For example, if your cam has .571-inch lift, you would measure the open loads at 1.50 inches (2.050 - .57 = 1.48). The 2.050 is the 6.4's detailed installed height. In production engines, if the max valve lift changes, so should the open load height.

In aftermarket, spring and cam manufacturers tend to select round numbers for open loads, such as .500- or .600-inch valve lift. Therefore, an aftermarket spring for the 6.4 might be rated at 1.45 inches (2.05 - .60).

Loads are best used to discover weak springs. For this selection process, open loads are better than closed loads. A valvespring selection process only works if you have more than 16 springs.

Spring Seats

The spring seats are machined into the aluminum cylinder head.

The 5.7 shims or seats are flat and about .045-inch thick. The 6.4 engines use a washer that has an upside-down "T" shape, with an .887-inch ID. Gen III Hemis use a spring seat OD of about 1.41 inches. The production spring OD is about 1.35 inches. Similar to the flat 5.7 shim, the 6.4 T-shim is about .043-inch thick.

Aluminum heads use a large, flat (and thin) washer, or shim, between the spring seat in the head and the bottom of the valvespring. There are two styles: flat (5.7) and upside-down T (6.4). Without a shim, the valvespring digs into the aluminum and destroys the spring seat and perhaps the head.

Valveguides

Valveguides are part of the cylinder head, but the valvespring must fit over the guide. The guide's OD and height above the spring seat are important, especially with dual valvesprings (beehives are standard). The guide height must allow the installed spring height and the valve lift that is defined by the cam and rocker arm. This is not much of a concern with beehive springs but can be an issue with a dual spring because the dual-spring ID tends to be less than that of the beehive spring.

As the valve lifts increase, the keeper-to-guide clearance can become an issue, which can be solved by cutting, or shortening, the guide (above the head).

Coil Bind

To measure coil bind, or solid height, the beehive spring is usually tested on a spring tester or in a vice. The spring is squeezed between the

The production beehive spring is quite tall, which is required by the 2.05-inch installed height. The free length (being measured here) is the opposite of coil bind or solid height.

two jaws of the vise or spring tester plates until it is solid. You measure the distance between the jaws or plates with a steel scale or a dial vernier. Only one spring needs to be tested of a given set.

If you have a dual (or triple) spring, it is typically disassembled, with the dampener and inner spring removed. Each piece is measured separately. The dampener is usually measured with the outer spring, but if the dampener is solid before the outer spring, it is usually trimmed (with tin snips).

Dampeners are usually designed by the manufacturer to have clearance at the outer spring's solid height. Along with load numbers, manufacturers usually list the solid height information.

Spring manufacturers often run performance springs closer to the coil bind. Solid height clearance used to be .100 inch, but today it's .050 to .075 inch. For example, if the valve lift is .675 inch and the installed height is 2.000 inches, the lift at max valve lift is 1.325 inches. If the solid height is 1.275 inches, you would have .050-inch clearance.

Beehive springs do not use dampeners; they are not required.

Valvetrain Upgrades

Cam lift dictates which spring is used, within the restrictions imposed by the installed height. Installing longer valves to achieve a 2.00-inch installed height can be expensive. These parts have limited availablility.

Package*	Best Cam Lift (inch)	Rocker Arm Upgrade	Valvespring Upgrade**	Installed Height (inches)
Stage 1, 2, and 3	.500	Stock	Stock	1.80
Stage 4 and 5	.550	T&D Street***	Comp Cams	1.80
Stage 5	.590	Stock	Stock	2.00
Stage 6 and 7	.600	T&D Street***	Stock or Comp Cams	2.00
Stage 8	.675	T&D Race	Comp Cams	2.00
* Package numbers refer to the list of "Performance Packages" in Chapter 6. ** All "best valvesprings" are beehive designs. *** T&D uses "street" to separate the two styles of rockers.				
The T&D rockers, both street and race, can be added to any package; they are optional.				

Retainers, Keepers and Seals

Keepers, or locks, are defined by the valvestem (grooves and stem diameter); the retainer is defined (or selected) by the valvespring (OD, ID, inner spring seat). They join at the angle of the keeper, 7 or 10 degrees. The retainer must match the keeper's angle.

Production keepers are 7-degree designs; the 10-degree examples are a race-only part. With aggressive high-lift race cams, the high-load springs required with these cams try to pull the 7-degree keepers through the retainer. The 10-degree keeper/retainer package solves this problem. I wouldn't consider this until you are in a position to use mechanical roller cam designs and lifts over .750 inch.

Retainers

The retainer attaches the valvespring to the valvestem, so it can control the valve's movements. The retainer must match the valvespring whether it's a single spring, dual spring, or triple spring. It also must match the keeper's back-angle, which is 7 or 10 degrees. It also must match the valvestem diameter, 5/16 inch on all Gen III engines. If you convert to

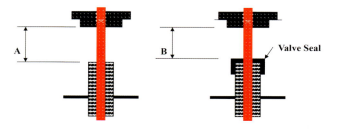

Retainer-to-Guide Clearance   **Retainer-to-Valve Seal Clearance**

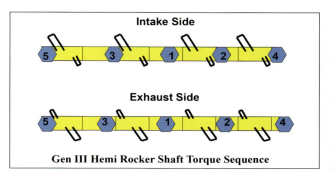

Gen III Hemi Rocker Shaft Torque Sequence

Even with a beehive spring, a valve seal must be used. The key clearance is "B" and this must be greater than the max valve lift plus .075 inch.

Prior to installing the rocker arms, rotate the crank 45 degrees from TDC so that all of the pistons are below the deck surface. Then torque the shafts to 195 in-lbs in sequence.

a dual-spring setup, 5/16-inch, dual spring retainers are available from the manufacturer.

Retainers are made from several materials; steel (chrome-moly) and titanium are the most popular. Race engine builders like titanium retainers. In many cases, the street engine builder wants to use steel retainers because they take a lot of abuse.

Because the top of the beehive spring is smaller than a straight spring, the retainer is lighter without changing materials.

Keepers

The keepers, or locks, must fit into the valvestem grooves and also match the stem size, which is 5/16 inch for all Gen III applications. Some keepers/valvestems have one groove and some have three grooves. The one-groove version is the most popular in performance engines (racing). The 6.4/392 engines use the three-groove version on both the intake and exhaust valves. Gen III grooves are round.

Seals

As discussed in Chapter 7, the valve seal must fit inside the valvespring; this makes the spring's ID very important. The trick with a valve seal is to leave enough room between the bottom of the keeper and the top of the valveguide so that the seal isn't crushed. The dual valvespring can be a challenge for any valve seal.

If your basic package does not have enough room for a seal (height), you should shorten the guide. Beehive springs tend to have more room inside the spring for the seal so it is less of a concern.

At .881-inch OD, this is a small retainer, which makes it very light.

The valve seal (to the left of the spark plug tower) fits onto the valveguide. They should be replaced when the head is removed and/or rebuilt.

Installed Spring Height

The valvespring must work with the cylinder head, valves, cam, and rocker arms; the key to all of this happening properly is the installed spring height. It is one of the most important measurements that you make during the engine building process. In some cases, the exhaust uses a slightly different installed height than the intake valve.

The installed height is defined as the distance between the spring seat in the head and the valvespring retainer (outer part) with the valve closed or on the seat. It is measured without the spring by pulling the retainer with keepers installed and the valve up against the valveseat and holding while the height is measured. It is typically measured with a snap gauge and a micrometer.

Everything about the head, valve, retainer, keepers, rocker arms, and cam are part of this number. The key is that the valve lift matches the valvespring's lift capacity, with a little or enough clearance, which is .050 to .075 inch today.

There are several common Mopar installed heights: 1.64 for the Magnum (small-block), 1.68 to 1.70 for the A-engine (small-block), 1.86 for the B/RB engines and the 426 Gen II (big-blocks), and 2.00 inches for long valves and racing applications. These springs generally do not "drop-in" to the Gen III applications because of their OD. The A-engine OD is around 1.46 and the big-block's is around 1.50, and the 2.00-inch hardware comes in around 1.58 or higher. The production head's spring seat diameter is about 1.41 inches.

A single spring and dampener has one installed height. The dampener height must be less than the spring height. A dual spring has two installed heights: one for each spring, inner and outer. Although the triple spring could have three installed heights, the inner two spring heights are usually the same; the retainer's seat width has to be greater to match. ■

There is plenty of room around the spring and retainer until the rockers and shafts are installed. A stock retainer is even smaller than the small end of a beehive spring.

After you remove the spring, re-install the keepers and retainer and pull the valve assembly against the seat. Use a snap gauge to measure the spring's installed height.

Once locked, measure the snap gauge height with a micrometer or dial vernier (2.070 inches here).

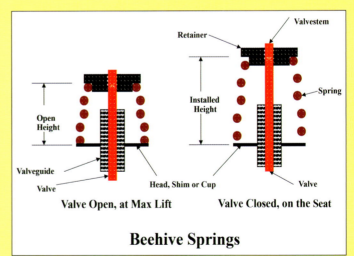

Beehive Springs

The installed height of a spring is just as important with beehive springs as it is with straight springs (single or dual). Valvespring manufacturers usually provide the spring specs, loads, heights, diameters, etc.

INTAKE MANIFOLD AND FUEL SYSTEM

Production sources and the aftermarket are still developing intake manifolds for Gen III Hemi engines. The production long-runner beer-barrel–style intakes do not lend themselves easily to the single- and dual-plane definitions and designs used on many previous V-8s. The aftermarket makes performance manifolds that can be used with either throttle body fuel injection or with carburetor(s).

Production Intake Manifolds

Gen III Hemi intake manifolds are unique in the typical, production V-8 engine group in that the manifold itself does not seal the tappet chamber and does not have to seal to the front and rear chinawall of the block, which is common in most small-block engines including the A-Engine, Magnum, and B/RB, and 426 Gen II big-blocks. There is no chinawall on either end of the block to seal to.

Part of the fallout of this major design change is that the tappets must be installed before the heads and the tappets can't be changed with the heads installed. The biggest advantage of this design is that it keeps the hot oil off the bottom of the intake manifolds.

All production Gen III Hemi intake manifolds are based on MPI. In the past few years, it has been popular to swap this engine into older cars, such as muscle cars, street rods, customs, etc., and in many cases convert the engine to a carburetor. A few manifolds were introduced to use a carburetor or they can be machined two different ways: one for each approach, that is, a carburetor or MPI (fuel injection).

The new player on the block is the fuel injection (FI) system, which is halfway between a carburetor and the MPI system, which I call TBI (throttle body injection) or EFI (electronic fuel injection). A computer controls the MPI system, but I discuss only the manifold and its hardware here (the electronics that control it are discussed in Chapter 10).

Gen III Hemi intake ports and the parting line between the intake manifold and the cylinder head is unique. No other engine's manifold comes close to this setup and can't be adapted easily. A selection of intake manifolds is available, but it is not unlimited. Some versions were made a few years ago and may not be currently

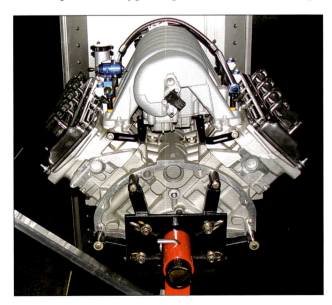

Production intake manifolds fit in the category of beer-barrel intakes. They have long runners coming straight out of the head and then curl over forming a large, round shape that looks like a beer barrel on its side. The aluminum 6.1 intake is an example and the long runner and the curl are easy to see in this rear view. The entry into the head is almost perfectly straight.

The traditional single-plane has long runners thanks to the very high-rise manifold used on the latest 426 Drag Pak. It can be used as a 4-barrel throttle body or with a carburetor. This example is from Arrow Racing Engines.

available. New ones are popping up faster than they can be tested.

All production intakes look like large, round beer kegs. Most of the production Gen III intake manifolds are made of plastic, which is a Nylon 6 material. The exception is the 6.1 engine, which used one made of cast aluminum. The 5.7 and 6.1 engines have the throttle body attached to the manifold inlet pointing straight ahead. The 6.4 and 6.2 engines have the throt-

tle body pointing at the driver-side headlight, or at about a 45-degree angle. The 6.1 and 6.4 intakes use a large cross-sectional area runner in the manifold; the original 5.7 uses a smaller runner to go with the smaller intake ports in the 5.7 heads.

The best of these intake manifolds is the aluminum 6.1 version (P5155462). It is about 8 hp better than the plastic versions. The fanciest one is the 6.4 version, which is considered to be an "active manifold." This means that it switches runner length from long for torque to short for more power. This is important with large-intake valves, large ports, and high-lift cams. Although the 6.1 is no longer available new or as a service part, thousands of these production manifolds are in the marketplace. Edelbrock has introduced its aluminum version, but no side-by-side comparisons were available as this book went to print. A possible source is Mopar Pro Shop; it has even more performance parts than Mopar does. Or try Indy Heads.

Aftermarket Intake Manifolds

Until a year or two ago, there weren't many aftermarket intake manifolds designed for the Gen III Hemi other than those offered by Mopar Performance/Chrysler. Most non-production and non-Chrysler intakes for the Gen III are for use with a carburetor, not MPI fuel injection.

A few years ago, ported intake manifolds were the best choice for Gen III engines because very few other options existed. This has recently started to change and special manifolds are beginning to pop up. This is partly due to Mopar/Chrysler's limited presence in the performance market, unlike a few years ago when they had an aggressive line-up

The 5.7 engines use a black plastic manifold similar to the 6.1, a long runner with a straight-ahead throttle body location.

The all aluminum 6.1 intake uses the same straight-ahead throttle body. It went out of production with the 6.1 engine. The new Edelbrock may fill the aluminum intake void.

The 80-mm stock throttle body attaches by four bolts and flows straight into the 6.1 manifold.

The 6.4 intake manifold is also black plastic, but the throttle body mounts at a 45-degree angle to the left rather than straight-ahead.

Edelbrock introduced its new beer-barrel intake for the Gen III engine, which is cast aluminum and painted black. The front throttle body accepts up to the 92-mm Hellcat throttle body. The top of the Edelbrock looks similar to the 6.1 part. Runner height and plenum size are not visible here. The injector holes and the fuel-rail attachments holes are visible. This manifold became available in the spring of 2017. (Photo Courtesy Edelbrock)

The 6.4 heads and manifolds have much larger intake ports than the earlier 5.7's. Modern Muscle offers adapter plates (lower left) to allow the use of the big-port intake with the small-port heads. The kit includes the longer screws (upper left) and the two styles of paper intake gaskets (right): one for the top of the plate and one for the bottom of the plate.

The finished machined dual-plane intake showing the throttle body mounted along with the fuel rails. Modern Muscle offers this dual-plane intake with or without fuel injection. The company also offers an adapter to turn the production single throttle body 90 degrees and bolt to the center flange. There is no float-bowl so it doesn't matter which direction the air comes from. (Photo Courtesy Modern Muscle)

of manifolds designed and built by Mopar/Chrysler.

In trying to decide where to discuss the various manifolds, I decided to put the Chrysler/Mopar non-production manifolds in with the aftermarket versions because they look like aftermarket manifolds and definitely do *not* look like production intakes.

The replacement intake manifold by Edelbrock is an amazing piece because it looks like the 6.1 cast-aluminum beer-keg production manifold, except that it is a little taller. It became available in April 2017. Early tests hint at 25 hp, but throttle body use and other aspects are still being worked out.

Manifold Types

All of the current production intake manifolds are classified as beer-barrel intakes. Due to the vertical mounting position, these production intakes cannot be used with carbu-retors. Dual-plane or single-plane as well as the number of carburetors or throttle bodies typically define aftermarket/performance intakes. Because the throttle body works in either the vertical or horizontal position, these aftermarket intakes can usually be used with either a carburetor or throttle body. However, machining differences at the manufacturer may require adapters.Dual-Plane 4-Barrel

Gen III production manifolds do not lend themselves to the traditional dual-plane/single-plane comparison. However, Modern Muscle offers a dual-plane intake for the Gen III. It is offered as a 4-barrel carbureted intake or as a fuel injected version with a large 1-barrel or 4-barrel throttle body that has the injector holes machined. It is made of cast aluminum. I recommend this intake for street rods, muscle cars, and customs that have used the Gen III engine in a swap. It should have good low-speed torque and good driveability. This dual-plane intake is designed to accept the stock single-bore throttle body for an easier swap based on production hardware.

Dual-Plane 8-Barrel

This intake is a true dual-plane. It is cast aluminum and made by Edelbrock. Earlier, Mopar sold a version of this intake (P5153556), but it is no longer available. This should be a good choice for the street 8-barrel look on the new high-tech engine in a muscle car or street rod. Edelbrock recommends two 500-cfm AVS carbs.

Single-Plane 4-Barrel

The single-plane intake has a central plenum with all eight runners coming into the single plenum. It tends to make more power at higher RPM than the typical dual-plane. On the flip side, it makes less torque at lower speeds. Because single-planes generally made more horsepower and ran more RPM, many people prefer them.

The aluminum single-plane for a 4-barrel carburetor use on the 5.7

One of the first carbureted intake manifolds for the Gen III (5.7 specifically) engine was this 8-barrel in-line intake by Edelbrock and Mopar. Still available from Edelbrock, it is a dual-plane version with much shorter runners than the production units. (Photo Courtesy Edelbrock)

This 8-barrel in-line carburetor system (two Edelbrock AFBs) is installed on a 392.

small-port heads is P4510582AB (carb). It makes more power (high RPM) than the production manifolds (aluminum and plastic) but hurts torque. Mopar no longer offers this version but you may be able to find one through Mopar Pro Shop. It's a good choice for the street rod and muscle car owner who is looking for more power and better performance at the strip.

Indy Heads offers a 4-barrel single-plane (carb and MPI). It is slightly taller than the other single-plane, but it has somewhat shorter runners and a bigger plenum. That may be another trade-off.

Single-Plane 4-Barrel (MPI)

This is basically the same intake as the carb version except that it is machined for fuel injectors at each cylinder. Mopar used to offer it (P4510582AB, MPI) for the 5.7 engine; today you can try Mopar Pro Shop. It's a good choice for the street rod or muscle car owner with a Gen III who wants the MPI technology with a retro look.

Single-Plane Race

In the lingo of intake manifolds, the single-plane manifold discussed

above is considered a standard riser. In the late 1960s and early 1970s, performance single-plane intakes were called high-risers because they were taller than a standard single-plane.

In this category of race single-planes, there are three heights, with no real way to distinguish one from the other so I created my own names: High-Rise, Xtra-High-Rise, and XXtra-High-Rise. Although they were designed for racing, they are not a race-only component.

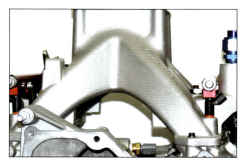

The XXtra-High-Rise single-plane intake is very tall and replaces the Xtra-High-Rise and 2-inch adapter on the 426 Drag Pak package.

Arrow Racing Engines makes several high-rise single-plane intakes so I made up names for them to help keep them straight: High-Rise, Xtra-High-Rise, and XXtra-High-Rise. This is the XXtra-High-Rise single-plane intake, using a Wilson large throttle body (the basic 426 Drag Rack) induction system. It is very tall and may not fit all engine compartments.

The mid-level high-rise from Arrow Racing, which I call the Xtra-High-Rise, isn't as tall as the XXtra version.

XXtra-High-Rise: This is the tallest of this group of tall intakes. It was designed for the 426 Drag Pak cars (with hood scoops) and is probably the best horsepower (high-end) single 4-barrel intake made to date. It is a cast-aluminum intake and was designed for MPI use; it has a very large throttle body. It is definitely a problem for hood clearance so it is probably not the first choice for cars without hood scoops.

It may be available from Arrow Racing Engines.

Xtra-High-Rise: This may be the second tallest intake for the Gen III Hemis and looks somewhat similar to the XXtra-High-Rise but is about 2.5 inches shorter. This was the original 426 Drag Pak intake (P5155184) and was used with a 2.5-inch spacer, billet and CNC-machined (from Arrow Racing). The XXtra-High-Rise was made as a one-piece manifold that didn't require the spacer. Therefore, if you do not use the spacer, it is much shorter and might fit in engine compartments that the tall manifold can't.

This manifold was offered by Mopar Pro Shop and Arrow Racing Engines but may no longer be available. It has been replaced by the XXtra-High-Rise. On any package concerned with hood clearance, use the High-Rise version.

The somewhat traditional High-Rise intake by Arrow Racing is the most interesting. It was introduced in late 2016 so it is too new for comparison tests, but it may be the best performance intake for these Gen III engines. It is low enough to fit most engines compartments and yet tall enough to have longer runners. By special machining, this intake can be used with a carburetor or with MPI and a 4-barrel throttle body.

High-Rise: This is intake is probably the best for street-strip use. It is offered in both carbureted and fuel-injected versions. It was designed to look like the standard, high-rise single-plane intakes used on small-blocks, but it stills looks like a member of this tall family. This manifold has slightly shorter runners than the XXtra-High-Rise version, but they are longer than the standard-rise version.

It also has large ports designed to work with the big-port 6.4 heads rather than the small-port 5.7. It is being used on the Canadian Pinty's circle-track series with a carburetor.

It is only available at Arrow Racing and comes in both configurations: carb and MPI.

Single-Plane 6- and 8-Barrel

Indy Heads offers a two-piece intake for Gen III engines that has multiple tops: one for a 6-barrel setup and one for an 8-barrel setup. It is a single-plane intake and has large runners for use with the 6.1 and 6.4 ports.

Indy Heads offers a single-plane short-runner intake that has some flexibility. Designed as a two-piece intake, the top is removable. The plenum is large, but the runners are short.

Indy Heads offers three different tops to seal the plenum on its single-plane intake: a single 4-barrel, a 3 x 2-barrel, and a 2 x 4-barrel.

Supercharged

The supercharged Hellcat 6.2 engine uses an IHI supercharger and its own intake manifold. Whipple, Magnuson, and Edelbrock offer kits (see Chapter 11 for more details).

The Hellcat has a manifold that is unique when compared to the other Gen III engines. Most aftermarket supercharger kits come with manifolds. Indy Heads offers a supercharger manifold for the

Rootes blowers, which are usually GMC 6-71 or 8-71 or clones such as those from Weiand and Blower Drive Service (BDS).

CNC-Ported

Most companies, including Modern Muscle, Indy Heads, and Modern Cylinder Head, that offer CNC-ported heads also offer CNC-ported stock intake manifolds. The one-piece manifolds have limits regarding how far up the porting can go. Aluminum manifolds are easier to install than the plastic versions.

Racing

At the time of this writing, drag cars in the Stock and SS categories use CNC-ported stock intakes where legal, and stock intakes where the rules limit modifications. Drag Pak cars with the 426 naturally aspirated package use Xtra-High-Rise or XXtra-High-Rise versions. Circle-track cars use the High-Rise.

Intake Manifold Upgrades

It is difficult to compare intake manifolds on the various Gen III engines. It used to be easy because you

only had a few choices: stock, ported, and 8-barrel. Today there are many options, with more in the works.

MPI Fuel System

The MPI system is used on all production Gen III Hemi engines. Multi-point injection means that there is at least one injector in each

This racing intake is a circle-track version, which means the Canadian circle track/road racing series. This is the approved single-plane (4-barrel carburetor) for this series. It is offered by Arrow Racing Engines.

port and the injector typically sprays on the backside of the intake valve. This also means that with an MPI system the throttle body has no fuel and only controls the airflow.

The other style of fuel injection is throttle body injection (TBI), or electronic fuel injection (EFI). With TBI, fuel is injected into the airflow in the throttle body. Instead of having eight

There is still a lot of interest in carburetors such as this vacuum-secondary Holley. The 650- and 750-cfm units are the most popular.

Manifold Alignment

Gen III Hemi intake manifolds have intake surface between the heads; it is horizontal as installed on the engine. It is not recommended that the head or deck surface be milled very much; .010- or .020-inch is the max because of issues with piston-to-head clearances, valve-to-piston clearances, and compression ratio. If this recommendation is followed, you should not have to mill the intake face on the head or manifold.

The exception to this guideline is if you use a shim or thick head gasket (.120 inch versus .040 inch) to get rid of too much compression; if you installed a supercharger on a stock 6.4 engine, for example. In this case the head moves

up .080 inch. In this case, I recommend moving the attaching bolt hole outward about .050 inch.

I also recommend grinding about .050 inch off the top of the intake runner and .050 inch off the bottom of the intake port in the head (head is removed).

The standard 5.7 head gasket is .027 inch and the 6.1 and 6.4 engines use .040-inch head gaskets. Cometic makes a thick intake gasket for use with the .120-inch-thick head gasket. This helps solve the "too much compression" problem that occurs most commonly when a supercharger kit is installed on a stock, production, naturally aspirated long-block, which comes with 10.2 or 10.7 CR. ■

MPI Intakes

Although some manifolds can be machined for both systems, I have listed everything separately. For MPI use, the ported intake made sense but ported manifolds for carb use would be a race part. The following is Upgrade Number 2.

Engine/Package	True Street	Upgrade Number 1	Street-Strip
2003–2008 5.7 Cars and Trucks	Stock	Ported*	†
2009-present 5.7 Cars and Trucks	Stock	Ported*	†
6.1	Stock	Ported†	†
2011 and newer 6.4	Stock	Ported*	High-Rise Single-Plane
Street Rod and Muscle Car with 5.7	Stock/Single-Plane†††	High-Rise Single-Plane†	
Street Rod and Muscle Car with 6.4	Dual-plane	High-Rise Single-Plane**	Xtra-High-Rise Single-Plane**
Off-road Trucks††	Dual-plane	High-Rise Single-Plane**	High-Rise Single-Plane**
410 ci and up	High-Rise	Xtra-High-Rise	XXtra-High-Rise Single-Plane***
Custom or Show§	Indy 6-Barrel	Indy 8-Barrel	XXtra-High-Rise Single-Plane

* Modern Muscle offers ported plastic intake manifolds.
** Upgrade assumes more power desired.
*** The engine wants the XXtra-High-Rise manifold, but hood clearance issues must be resolved.
† The High-Rise intake (Arrow Racing) is too new to compare high-performance versus ported stock and 5.7 heads.
†† Off-road Trucks and Jeeps want max torque and driveability, small ports sealing and flow versus large ports in a manifold.
††† The ported stock manifold may make more power, but the single-plane may be desired for appearance; a high-performance comparison is not available yet.
§ Selection is based on appearance only, not on performance. The Edelbrock 8-barrel would be an equal to the Indy 8-barrel. The XXtra-High-Rise was selected for Street-Strip because the height makes it look impressive and it would accept new fuel injection systems (plus, the Xtra-High-Rise is no longer readily available).

Most of these manifolds can be obtained in either carbureted or fuel-injected versions, such as the dual-plane (Modern Muscle) or the High-Rise (Arrow Racing). Obviously, stock and ported manifolds must be used with MPI. Throttle body versus carburetor is another decision you need to make.

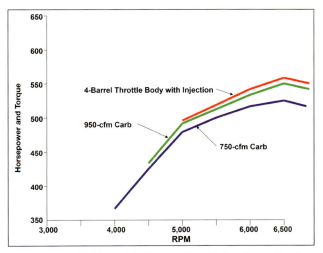

Bigger engines need more airflow so a carburetor/throttle-body test is in order. I selected a 392-ci engine for the baseline. I also used a single-plane intake (from Mopar Performance). The High-Rise (from Arrow) and the dual-plane (from Modern Muscle) were not available at test time. The basic comparison was to use the basic 750 4-barrel carb (from Holley or Quick Fuel) as the baseline and the 950-cfm unit (from Holley or Quick Fuel), which made about 10 hp more. The fuel injection used a 1,000-cfm 4-barrel throttle body (from Edelbrock or Fuel Air Spark Technology [FAST]), which made about 5 hp more than the big carb. Using a high-rise manifold (arrow) would have made these numbers much higher.

injectors, this system uses two or four injectors. It has no fuel rails or complicated wiring harnesses.

MPI is an easy system to work with if the factory MPI system is your baseline because it is used on all Gen III Hemis. It is more complicated if you are making a conversion or a custom installation.

Independent of the basic hardware (intake manifold, throttle body, fuel rail, eight injectors, and electric pump in the tank), six to eight sensors must be wired in and operating for the ECM to know what's happening with the engine so everything works properly. The production Gen III system has eight sensors that feed information to the computer. Aftermarket

Intakes for a Carburetor

Although the chart below is very similar to the one for MPI use, the details are unique because stock Gen III manifolds cannot be used with a carburetor. The following is Upgrade Number 2.

Engine/Package	True Street	Upgrade Number 1	Street-Strip
2003–2008 5.7 Cars and Trucks	4-Barrel Dual-Plane	High-Rise Single-Plane	Xtra-High-Rise Single-Plane*
2009-present 5.7 Cars and Trucks	4-Barrel Dual-Plane	High-Rise Single-Plane	Xtra-High-Rise Single-Plane*
6.1	4-Barrel Dual-Plane	High-Rise Single-Plane	Xtra-High-Rise Single-Plane*
2011 and newer 6.4	4-Barrel Dual-Plane	High-Rise Single-Plane	Xtra-High-Rise Single-Plane*
Street Rod and Muscle Car with 5.7	4-barrel Single-Plane	4-Barrel Dual-Plane	Xtra-High-Rise Single-Plane
Street Rod and Muscle Car with 6.4	4-Barrel Dual-Plane	High-Rise Single-Plane	Xtra-High-Rise Single-Plane*
Off-road trucks/Jeeps†	4-Barrel Dual-Plane	High-Rise Single-Plane**	Xtra-High-Rise Single-Plane**
410 cc and up	Xtra-High-Rise Single-Plane	Xtra-High-Rise Single-Plane	Xtra-High-Rise Single-Plane
Custom or Show††	Indy 6-Barrel	Indy 8-Barrel	XXtra-High-Rise Single-Plane

* The Xtra-High-Rise (from Arrow Racing) is tall and may have hood clearance issues. The XXtra-High-Rise has replaced it. Use the High-Rise to solve hood-clearance issues.
** Upgrade assumes more power is desired. Dual-planes make the best torque.
† Off-road trucks and Jeeps want max torque and driveability.
†† Selection is based on appearance only, not on performance. The Edelbrock 8-barrel would be an equal to the Indy 8-barrel.

The XXtra-High-Rise was selected for Street-Strip because the height makes it look impressive and it would accept new fuel injection systems (plus, the Xtra-High-Rise is no longer readily available).
Throttle body upgrades and carburetor upgrades are discussed later in this chapter.

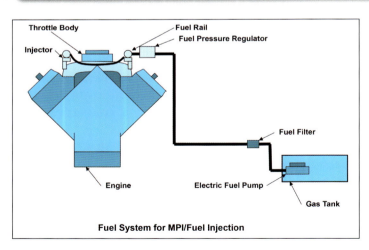

Fuel System for MPI/Fuel Injection

If you are doing an engine swap with an MPI engine, this is all the hardware that you need.

MPI units use a similar number.

The brake specific fuel consumption (BSFC) is the ratio of fuel consumed (pounds per hour) to the horsepower produced. Most OEM or factory gasoline engines have a BSFC of about .50. Highly tuned normally aspirated race engines operate in the .40 to .45 area. Typical turbocharged or supercharged engines run in the .55- to .60-BSFC range. For engines using methanol for fuel, the BSFC factor is doubled: 1.00 for naturally aspirated versions and 1.10 to 1.20 for supercharged engines on methanol.

The most important aspect of the fuel that you use is its octane rating. Race cars use race gas with octane ratings over 100. Street/Strip cars are best off if they focus on pump gas, or 92 octane premium. See Chapter 7 for more information.

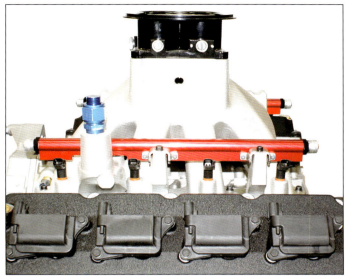

A throttle body injection system injects the fuel into the throttle body; an MPI system injects the fuel directly into each port, typically at the head-to-manifold interface. The fuel rail sits on top of the four individual injectors. There is no fuel in the 4-barrel throttle body mounted on top.

Throttle Bodies

Most production Gen III throttle bodies have a large, round, single throttle bore. The typical size is 80 mm, which is pretty big, but there are bigger versions. Most upgrades today are based on the single, large, round throttle body, but as the High-Rise manifolds become more available, this may change to the large 4-barrel throttle bodies, which are already available.

Single Round

The production throttle body is an 80-mm unit. Available from

Sensors and Injectors

All Gen III Hemi production engines use multi-point injection (MPI), which is very high tech and requires numerous sensors in addition to fuel-injection hardware. These production systems require 10 sensors, but the speed sensor and knock sensor(s) are not usually included. With all of these sensors, the engine controller, or computer, can generate extremely accurate fuel and spark maps for the engine controls.

Crank Position Sensor

This sensor is located toward the rear, on the passenger's side of the block at the rear edge of the number-8 cylinder. It counts the number of teeth on the crank wheel: 32 or 58.

Cam Position Sensor

This sensor is located on the side of the front cover.

Engine Coolant Sensor

This sensor is located next to the thermostat housing, on the top front of the front cover/water pump. It tells the computer the temperature of the engine so it can adjust the fuel level and spark advance.

Oxygen Sensor

It measures the amount of oxygen in the exhaust gas. Sometimes there is only one, but newer models may have two or three of these along the exhaust pipe. These sensors are heated.

Air Temperature Sensor

This sensor is installed in the intake manifold and typically in one of the front runners.

Manifold Absolute Pressure Sensor

Commonly known as the MAP sensor, it is located in the front of the throttle body. This sensor measures the pressure inside the manifold as the engine load varies.

Throttle Position Sensor

Often referred to as the TPS, it is located on the side of the throttle body. It is a variable resistor, which feeds the throttle position information to the computer.

Idle Air Control Motor

Called the IAC motor, it is located on the throttle body. It adjusts the engine idle speeds. Never attempt to adjust the engine idle speed by using the screw on the IAC motor or linkage.

Speed Sensor

This sensor is located in the transmission extension and is not considered an MPI sensor because it is used to tell the speedometer how fast the vehicle is going; it is not directly related to fuel or spark.

Knock Sensor

There are two of these: one on each side of the block. They are not part of the MPI system but pull spark advance if they sense knock (detonation). They are extremely important for supercharged engine applications and any engine using pump gas.

Fuel Injector

There are eight injectors in each Gen III engine package. The production injector changes with the engine's horsepower output so there is one for the 385-hp 5.7, one

Sensors and Injectors *CONTINUED*

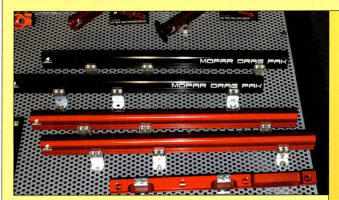

The fuel rail is different on a car and a truck because it is designed to fit the body style, model, etc. The exception is the billet fuel rail used on the 426 crate engine and the Drag Pak engines (available from Arrow Racing Engines). It has a larger ID than the production fuel rail, so it holds more fuel and flows more fuel, which is better for high-horsepower packages.

for the 485-hp 6.4, and one for the 707-hp 6.2 Hellcat. The injector must fit into the machined hole in the intake manifold and into the fuel rail, and it must have the right connections for the fuel-control wiring harness. ■

Injector Size Calculation

Injectors are rated by the amount of fuel that they flow. Injectors must be sized for the revised performance of the engine (horsepower). Complicating this issue today is that some injectors are flowed, or rated, at 4 bar pressure in addition to the 3.0 and 3.8 bar pressures more commonly used.

To calculate the proper injector size for your engine you can use the following formula:

$$HP = FR \times N \times .8 \div BSFC$$

Where:

HP = horsepower potential of the engine
FR = flow rate of the injector at 3.8 bar (pressure)
N = number of cylinders
.8 = a practical maximum injector flow rate for street use
BSFC = .5 for a typical V-8 engine

For example, a 5.7 engine making about 450 hp, the equation is as follows:

$$38 \times 8 \times .8 \div .5 = 486 \text{ hp}$$

Therefore, the stock injectors are adequate, but you should upgrade to 45-pound injectors if you reach 500 hp.

Modern Muscle are ported 80-mm units, which flow a lot more air and make a very nice first upgrade in the induction area. Several 84-mm units are available; an 88-mm unit is available from BBK. Edelbrock makes a 90-mm throttle body, but it does not have the Gen III hardware yet. The Hellcat has a 92-mm throttle body and there are 95-mm upgrades, but these might be too large for stock manifolds.

4-Barrel

In the late 1990s, Mopar Performance offered an aluminum 4-barrel

BBK offers a larger throttle body (up to 88 mm) single-bore.

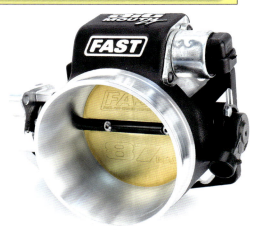

FAST offers larger single-bore throttle bodies. This one is up to 87 mm, with bigger optional units. (Photo Courtesy Fuel Air Spark Technology)

There are many suppliers for the standard 4-barrel throttle body. They flow about 1,000 cfm and use 1¾-inch throttle bores. Edelbrock offers this one.

throttle body. It was designed to fit instead of the standard Holley 4-barrel carburetor that was popular at the time. It features 1¾-inch throttle bores and flows about 1,000 cfm. No longer offered by Mopar Performance, several companies make them, including Edelbrock.

Large 4-Barrel

These giant throttle bodies are based on the Holley 4500 carburetor's bolt pattern. The somewhat standard unit is the 2-inch throttle bores that is used on Hemi Drag Pak

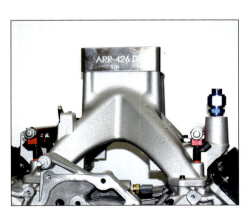

The aluminum adapter mounted on top of the Xtra-High-Rise manifold adapts the 4500-style throttle body to the standard Holley 4-barrel intake manifold. Without the taper in the adapter the higher flow of the bigger throttle body is lost. A 2-inch adapter is not used on the XXtra-High-Rise intake in the 426 Drag Pak. The XXtra-High-Rise intake replaces the earlier Xtra-High-Rise that used the 2-inch adapter. The tapered aluminum adapter and spacer adapt the big 4500 carb-bolt pattern to the standard Holley 4-barrel pattern. It works with both large throttle bodies and carburetors. This example is from Arrow Racing Engines.

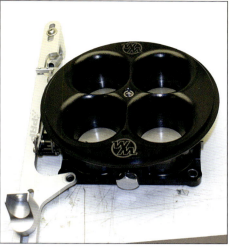

Wilson Manifolds offers some very large throttle bodies. This one uses four 2-inch throttle bores and flows about 1,600 cfm; several versions are available. This unit was used on the 426 Drag Pack engine with the Xtra-High-Rise and XXtra-High-Rise intakes. Wilson also offers even larger throttle bodies, and it has a unit that has 2.35-inch throttle bore and flows about 2,300 cfm.

cars. This is the throttle body that the XXtra-High-Rise intake manifold was designed for and it was also used on the Xtra-High-Rise intake with a thick spacer. Wilson makes the throttle body; it is also available from Arrow Racing Engines.

There are two carb/throttle body-attaching patterns: the standard Holley pattern, or 4150, and the bigger 4500. Throttle bodies reflect the carburetor pattern. To adapt one size to another, you need a tapered adapter; these are from Arrow Racing Engines.

EFI System

The EFI systems are more easily understood if they are referred to as TBI to separate them from the production-style MPI system. These systems have no injectors in the intake runners or the cylinder head. The fuel injection takes place in the throttle body. EFI sounds more technical and seems to have become the hot new product. Although some look like a carburetor, others look unique. In general, I would match the flow to the carbs in the "Carburetor Upgrades" sidebar and use the same intake manifolds.

Edelbrock offers an EFI system called the E Street 2 EFI, which can

These units are parts of the Edelbrock Pro-Flo EFI package, which is complete with lots of parts. (Photo Courtesy Edelbrock)

Throttle Body Upgrades

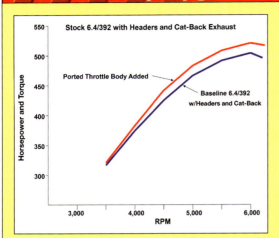

The first three lines of the chart below assume production vehicles so only MPI systems are included.

With induction systems, you have to be careful not to overdo it. Installing a giant throttle body on a street engine would not be a good choice. Although not flashy, I selected a more useful comparison. I chose a 6.4/392 engine for the baseline with headers and a cat-back exhaust (520-hp baseline). My theory was that the owner wouldn't modify the intake system until he had done the exhaust, or perhaps at the same time. To this baseline, a ported stock throttle body was added and it gained 14 hp. The best thing about this approach is that everything bolts up and plugs in. I would like to have added the Hellcat 92-mm throttle body to this comparison but it wasn't available at test time.

Both Quick Fuel and Holley offer bigger carbs based on the standard Holley bolt pattern. This Quick Fuel unit is rated at 950 cfm. (Photo Courtesy Quick Fuel)

For MPIs				
Refer to "Performance Packages" in Chapter 6 for more details.				
Engine/Package	Head	Best Intake	Best Throttle Body	Bigger Throttle Body
5.7 (early)	5.7 (early)	Ported	Ported	80 to 88 mm
All 5.7	5.7 Eagle	Ported	Ported	80 to 88 mm
		Arrow Single-Plane*	4-Barrel	4-Barrel**
6.1 & 6.4/392	6.4 Apache	Ported	Ported	80 to 88 mm
		Arrow Single-Plane*	4-Barrel	4-Barrel**
Street Rods	all 5.7s	Stock	80 to 85 mm†	90 to 95 mm
	5.7 (early)	Modern Muscle Dual-Plane***	4-Barrel	4-Barrel**
	5.7 Eagle	Modern Muscle Dual-Plane***	4-Barrel	4-Barrel**
	6.4 Apache	Arrow Single-Plane*	4-Barrel	4-Barrel**
Muscle Cars	all 5.7s	Stock	80 to 85 mm†	80 to 88 mm
	5.7	Modern Muscle Dual-Plane***	4-Barrel	4-Barrel**
	6.4 Apache	Stock	80 to 85 mm†	90 to 95 mm
	6.4 Apache	Arrow Single-Plane*	4-Barrel	4-Barrel**
Customs	Eagle or Apache	Modern Muscle Dual-Plane***	4-Barrel	4-Barrel**
	Eagle or Apache	Indy 6-Barrel††	3 x 2-Barrel	4500
	Eagle or Apache	Indy 8-Barrel††	2 x 4-Barrel	4500

* The Arrow single-plane (or High-Rise single-plane) with 4-barrel throttle body is optional; too new for comparisons but appears to make more power.

** The standard 4-barrel throttle body is a 4 x 1.75, which flows about 1,000 cfm. There are larger ones (about 1,350 cfm) that bolt to the same 4150 attaching pattern and use the same intake.

*** The Modern Muscle dual-plane (also dual-plane 4-barrel) offers the best driveability.

† The stock manifold is least expensive and the 80-mm throttle body, ported or an oversized unit up to 85 mm, matches well.

†† The 3 x 2-barrel and 8-barrel intakes by Indy Heads and 8-barrel single-plane intake by Edelbrock and Mopar.

All injectors must be matched to the power package.

One of the new EFI producers is FI Tech, which offers several versions. They are self-contained units that bolt onto the standard carburetor flange/intake manifold. They are about the size of the typical AFB/AVS carburetor. They are self-contained and seem to be easy to use. The base unit is called the GoStreet EFI and supports up to 400 hp. The Go EFI 4 is good for 600 hp. The Go EFI 8 is good for 1,200 hp.

Several Holley EFI systems are available, from the Sniper EFI system with 650- and 1,250-hp versions to the Terminator Stealth EFI, which looks like a carburetor. (Photo Courtesy Holley)

The FAST EZ-EFI is a complete kit. The fuel ECM can be used without the TBI unit and combined with the ignition module for a complete MPI system. (Photo Courtesy Fuel Air Spark Technology)

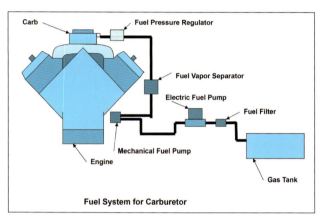

Fuel System for Carburetor

The carburetor fuel delivery system could also be used for the most TBI/EFI systems, but some TBI systems are designed to work with the fuel-pump-in-tank MPI system or other variations. Consult the manufacturer concerning fuel delivery details.

Fuel Vapor Separator
(mount vertically)

The vapor separator is a special fuel filter that is designed to keep the engine from vapor locking. It is not used on MPI systems but I recommend it for use with carburetors, especially in hot weather and in a cruise situation. It was designed originally for the 426 Gen II Hemi and the 440 6-barrel and is currently offered by Year One.

support about 600 hp. The Pro-Flo EFI system supports up to 775 hp.

Carburetion

Carburetors were the standard fuel-metering unit in the 1960s and 1970s through the muscle car era. The big question was: Holley or Carter? In the late 1970s and 1980s the TBI system came into production. MPI injection systems began in 1984 and were everywhere by 1993. Carburetors dropped out of production completely sometime in the late 1980s.

Based on that information you would think that only fuel injection options would be of interest today. However, that is not the case. Owners want carburetors for their hot rod projects. Intake manufacturers make intake manifolds that use carburetors for engines that came with fuel injection, which includes the Gen III Hemis.

Carburetor Performance Options

This chart is for carburetors; it is similar to the earlier chart for MPI systems. Refer to "Performance Packages" in Chapter 6 for more details. ∎

Carburetor or throttle body spacers or adapters come in many sizes, such as .500, 1.00, and 2 inches, and in several materials, including plastic and phenolic (upper right). My favorite is the phenolic for general street use. (Photo Courtesy Hedman)

Carburetor/Intake Upgrades				
Engine/Package	Head Used	Best Intake	Best Carb Size*	Upgrade**
5.7 early	5.7 early	4-Barrel Single-Plane	600 to 650 cfm	750 cfm
All 5.7	5.7 Eagle	4-Barrel Single-Plane	750 cfm	800 to 850 cfm
		Arrow Single-Plane	750 cfm	800 to 950 cfm
6.1 & 6.4/392	6.4 Apache	Arrow Single-Plane	750 cfm	800 to 950 cfm
		Arrow Single-Plane	800 to 950 cfm	4500 w/adapter
Street Rods	all 5.7s	Modern Muscle Dual-Plane	650 to 750 cfm	800 to 850 cfm
	5.7 Eagle	Modern Muscle Dual-Plane	750 cfm	800 to 950 cfm
	6.4 Apache	Arrow Single-Plane	750 to 950 cfm	4500 with adapter
Muscle Cars	all 5.7s	Modern Muscle Dual-Plane	650 to 750 cfm	800 to 850 cfm
	6.4 Apache	Modern Muscle Dual-Plane	750 to 950 cfm	4500 with adapter
	6.4 Apache	Arrow Single-Plane	750 to 950 cfm	4500 with adapter
Customs	Eagle or Apache	Modern Muscle Dual-Plane	600 to 950 cfm	4500 with adapter
	Eagle or Apache	Indy 6-Barrel	3 x 2-Barrel carbs	3 x 2-Barrel carbs
	Eagle or Apache	Indy 8-Barrel	2 x 4-Barrel carbs	2 x 4-Barrel carbs
	Eagle or Apache	8-Barrel Dual-Plane	2 x 4-Barrel 600 cfm	2 750-cfm carbs

* The extra-large 4500 family of carburetors can be used on the Arrow single-plane by using a Wilson adapter (from Wilson or Arrow Racing).
** The 950-cfm carb is available from Holley and Quick Fuel; the 800/850-cfm AFB/AVS carb is available from Edelbrock.

Carburetor Selection

More than 3.5 million Gen III Hemis have been built, and there are many places to use them besides production cars and trucks. I would not expect an owner to remove the production MPI system and install a carburetor on a new production car or truck. However, if that owner installs the engine into a 1970 Duster or a 1968 Road Runner, he or she might be interested in a carburetor.

Carb Size

A variety of carburetor sizes can be theoretically calculated, but in real-ity only some are actually available. For street use, you would like to use a carburetor with vacuum-secondaries (from Holley or Quick Fuel) or an AVS version (from Edelbrock). Over the past few years, Quick Fuel and Holley have offered higher-flow 4150 pad carburetors that are in the

Determine Carb Size for Displacement

It is always interesting to use a formula to predict carburetor size and engine displacement.

$$CCFM = D \times MRPM \div 3{,}456$$

Where:
CCFM = carburetor cfm
D = displacement of the engine (ci)
MRPM = engine's maximum RPM
3,456 = Mathematical constant

For example, if your engine is a 5.7 V-8 (347 ci) and you use a hydraulic roller cam (about 6,000 rpm on the street), your carb CFM should be 602 cfm (347 x 6,000 ÷ 3,456). The small AFB and AVS 4-barrel production carbs were probably about 600 to 650 cfm.

If your engine is a stroker with 426 ci and a big hydraulic roller cam (about 7,000 rpm), your carb CFM should be 863 (426 x 7,000 ÷ 3,456).

900- to 1,000-cfm range and bolt to the standard manifold opening. These new carbs typically do not offer vacuum secondaries because they were not built for the street. Quick Fuel and Holley have also built smaller versions of the big 4500 carburetor that offers 950 and fewer CFM.

Quick Fuel and Holley carburetors use one attaching bolt pattern while the Carter-Edelbrock carburetors use a slightly different pattern. In most aftermarket manifolds, both patterns are machined in. Adapters that are about 3/4-inch thick offer the ability to switch from one pattern to the other.

There are basically three carburetor attaching patterns: the standard 4150 Holley 4-barrel pattern, the similar standard Carter/Edelbrock pattern, and the giant 4500 pattern, which is based on the Holley 4500 race carburetor.

Intake Gaskets

The type of intake gasket is dictated somewhat by the intake manifold. The 6.1 aluminum manifold uses a steel and rubber intake gasket; the 6.4 uses a rubber O-ring in the groove in the manifold. Although the O-ring might be better in some cases, it requires that the manifold have the groove in it, which could be expensive to add.

Air Cleaners

The air cleaner is dictated by hood clearance and scoop availability. The Gen III Hemi has used many hoods and hood scoops. The 426 Drag Pak has a unique air cleaner (available from Arrow Racing Engines). The Indy 6-barrel intake can use the same base plate as the production 440 6-barrel from 1969 to 1971. The Edelbrock and Indy Heads 8-barrel intake systems should be able to use the 426 Street Hemi inline 8-barrel base plate, but check with the intake manufacturer to be sure. With multiple carbs, one common air cleaner is better than two or three smaller air cleaners.

Hood Scoops

Many different hood scoops are used on Gen III Hemi cars and trucks. Most look like earlier models from the muscle car era: the 1969 440 6-barrel, the Challenger, T/A, etc. The hood scoop has two basic functions: to offer increased hood clearance to the throttle body or carburetor and to supply cold air to the engine. You have lots of options here.

Cold Air

High-performance air cleaners relocate the air inlet from the top of the engine to below the radiator/support to pick up cold air from below the radiator that is passing below the bumper. Cold-air systems generally come with a low-restriction air cleaner as part of the package. The relocated air box is typically connected to the throttle body by a large-diameter smooth tube with gentle bends designed for high flow.

Several aftermarket companies offer cold-air or fresh-air systems that replace the stock air cleaner with a high-flow system; this one is from Mopar Performance.

IGNITION SYSTEM

All Gen III Hemis have MPI, which controls both the ignition and the fuel. I discussed the fuel hardware in Chapter 9, but here I discuss the actual ECM for the ignition system.

Brief Overview

The MPI's ignition system is made up of many parts and most of them are outside the engine except the sensors, which attach to engine components such as the throttle body, intake manifold, water pump, etc. The MPI ignition's job is focused

Many aftermarket suppliers offer ignition boxes for the Gen III Hemi. This one is from FAST and features two physical boxes: one is for the fuel side and one is for the ignition side. You can split it if you use a carburetor for your high-tech ignition system.

inside the Hemi combustion chamber: to fire the spark plugs. All Gen III Hemis use dual-ignition plugs. This feature was first used in racing in the early 1970s in the Gen II Hemi.

The ignition system is required to ignite the air/fuel mixture inside the combustion chambers of the engine. The ignition parts are the spark plugs (2), ECM, coils, switches, wiring, battery, starter, ground straps, and at least six sensors to tell the computer what the engine is doing.

With the introduction of fuel injection (MPI), the ECM controls both the ignition functions and the fuel functions. With the Gen III MPI system there is no distributor and no plug wires (coils attach directly to the plugs) and there are two plugs per cylinder. The six sensors allow this to occur. It takes basically the same six sensors to tell the computer what it needs to know to control the fuel demand for the engine; so once you have one, adding the second engine parameter is easy.

Most Gen III engines since 2004 have MDS and most 6.4 engines use VVT, but neither directly affects the ignition system, although the same computer controls both (see Chapter 2 for more information).

With the MPI system, the ECM controls everything and knows everything about the engine. This allows the engine to run much higher compression ratios on the same pump gas that was available years ago with the carburetor and TBI systems. It allows higher engine outputs while still meeting strict emissions levels. Properly maintained, it allows for even more durability and reliability from the engine. This leading-edge technology is great for engine performance, but it has a down side.

In Chapter 9, I discussed the fuel concept of BSFC, which basically stays the same for the engine except for superchargers and alcohol fuel. This means that if you want to make more power, you must provide more fuel. The computer has a small window for performance upgrades and after that, it must be reprogrammed, or re-flashed. This is related to the fuel curve not the spark/ignition curve.

However, Gen III Hemi engines have two knock sensors (detonation sensors) that are not counted in the six computer inputs. These knock sensors pull back the spark (smaller numbers) if it senses any detonation; in other words, it saves the engine from damage. As the engine's output

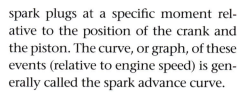

Ignition Specifications		
	5.7, 6.1, 6.4, 6.2, 426	Single-Plug Option*
	ECM/Computer**	ECU***
	Distributorless	*Distributor Added†*
Firing Order	1-8-4-3-6-5-7-2	1-8-4-3-6-5-7-2
Spark Plug	Dual-Plug	Single-Plug + Blank
Plug Wires	None (1)	Long-Boots Required††
Total Timing	23 to 25 Degrees	35 Degrees†††

* With Gen III MPI, the coils mount directly to the plugs; no plug wires.
** The MPI system uses an ECM plus six sensors to control the ignition and fuel for the engine; no distributor.
*** The single plug was used with a distributor and the computers were called ECUs.
† The distributor was used in the single-plug configuration.
†† Once you switch to a distributor, spark plug wires are needed with long boots.
††† When the distributor conversion is made and switched to a single-plug configuration, timing increases to 35 degrees.

is increased, the ECM needs to be replaced or reprogrammed. There are many components to reprogramming.

Advance Curve

With the new engines, there is still an advance curve, but it is inside the computer. The various sensors tell the computer what the engine is doing and it figures out the spark advance based on the curve built into the computer, for all three aspects of the advance: initial, centrifugal, and vacuum. The computer also indicates which coil/plug should fire. The distributor performed all of these functions. If the ECM is reprogrammed, the programming changes the initial setting or the advance curve.

Spark Advance

The engine's spark advance has three aspects: initial advance, centrifugal advance, and vacuum advance. With no distributor, all of these functions are performed inside the computer. The key to the engine making torque and horsepower is to fire the

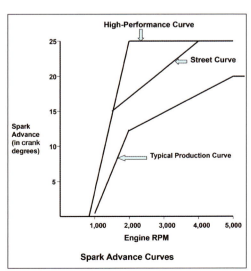

Spark Advance Curves

All of the spark advance curve is now done in the ECM. No more springs, weights, or vacuum lines. The production curve is an over-simplified version of what comes with a stock engine. The street curve is basically what you want today using pump gas. The high-performance curve is used in racing with good gas. Moving, or changing, this curve is one of the items that "reprogramming" does to your ECM's ignition map. In practice, reprogramming affects, or adjusts, the fuel side more than the spark side.

spark plugs at a specific moment relative to the position of the crank and the piston. The curve, or graph, of these events (relative to engine speed) is generally called the spark advance curve.

Initial advance is the number that you used to set (with a timing light) by rotating the distributor housing while the engine was at idle. Now it is programmed into the computer.

The centrifugal advance used to be set by a spring-and-weight mechanism inside the distributor housing and increased the advance with engine speed (RPM). This advance change was controlled by two small springs. Re-curving the distributor, a common performance trick, was accomplished by changing the springs. Now, all this is done by an equation inside the computer.

The third type of advance is vacuum advance, which is an add-on advance. It was basically a fuel economy and drivability assist. It is important to a street car but was often disconnected (plugged) for racing. It too is done inside the computer.

The sum of the initial advance and the centrifugal advance is called the total advance. For racing applications you set the total advance rather than the initial advance. If the spark advance is graphed, or plotted, against the engine speed, or RPM, the resulting graph is called the spark advance curve.

Spark advance is much more critical on street engines or dual-purpose engines because they use the full range of engine speeds and loads; race engines tend to run at high speeds and a fixed advance. These types of advance are controlled inside the computer so to change any of these aspects today, the ECM must be reprogrammed. Typically, the vacuum aspect is not changed, only the total number.

For example, if the computer's centrifugal advance equation has 12 degrees built in and the initial advance is set at 8 degrees, you have a total advance of 20 degrees (8 + 12). If you increase the initial advance by 5 degrees, to 13 degrees, your total advance increases to 25 degrees (13 + 12).

Perhaps a more common reprogramming approach would be to increase the centrifugal advance by 5 degrees and leave the initial advance alone (8 + 17 = 25). Caution: This assumes that the engine needs more total spark advance: fuel quality, compression ratio, boost pressures if used, and so on.

Vacuum advance can be considered an add-on. The vacuum control used to be added to the side of the distributor so an add-on seemed appropriate. Now it is built in like the other aspects, so it is just part of the whole ignition map.

I'll continue with the above example. If the engine has 15 inches of vacuum at cruising speed (highway), total advance at this RPM (about 1,500 rpm with today's 6- and 8-speed transmissions) might be 30 degrees (8 degrees of the initial plus half of the centrifugal, 7) plus the 15 degrees of vacuum advance for 30 degrees (8 + 7 + 15) at cruising speed; better fuel economy.

If you go to wide-open-throttle, the vacuum drops to zero and the vacuum-advance also drops to zero, it's back to 15 (8 + 7) total advance until the RPM changes and it follows suit.

Crank Position Sensor

One of the most important sensors to the MPI computer is the crank position sensor. It is used with the MPI system only. It is mounted to

Compatible Crank Wheel Sensors			
Model Year	Engine Application	Crank Wheel Tooth Count	Part Number
2003–2006	5.7 Hemi and 2006 6.1	32	56028815AA
2007–2010	5.7 and 6.1 Hemi	32	05149009AB
2009–2012	5.7 (VCT) and 6.4 Hemi	58	05149230AA
2013–2016	5.7 and 6.4 Hemi	58 radius tooth	68140678AB

The one on the left is the 58-tooth sensor and the one on the right is the 32-tooth sensor. They must not be mixed or swapped.

the passenger's side of the cylinder block toward the rear, just behind the number-8 cylinder. It is positioned to look at the crank wheel, which is inside the crankcase and bolted to the rear of the last counterweight on the crank. There are four crank position sensors and three crank wheels. These sensors should be maintained (do not swap) because the computer must receive the proper signal from the crank position sensor.

The crank wheel is bolted to the crank. There are three: one with 32 teeth and two with 58 teeth. (See Chapter 3 for more details.)

Sensors

The production Gen III MPI system has six sensors that feed information to the ECM; the computer uses this information to make the ignition and fuel maps. The aftermarket MPI systems use a similar number. There are two position sensors (crank and cam) and two knock sensors plus the speed sensor that have inputs to the computer but aren't counted in the basic six.

The six MPI sensors are: engine coolant sensor, oxygen sensor, air temperature sensor, MAP sensor, throttle position sensor, and idle air control motor. The basic functions of these sensors are discussed with the fuel map in Chapter 9.

The oxygen sensor is located in the exhaust pipe(s) and there may be more than one. They measure the amount of oxygen in the exhaust gases during engine operation. This is important to spark advance and detonation (knock).

The two knock sensors are important to the overall operation of the engine because they allow the engine to operate in less-than-ideal conditions such as too much compression, low-octane gas, or superchargers with more than 10:1 compression ratios. The knock sensor decreases the spark advance when detonation occurs so the engine does not damage itself.

Computer

If you install a bigger cam, bigger valves, big port heads, ported heads, more cubic inches, headers, or a better intake manifold/throttle body, you need to reprogram the computer. The production computer has a window for modifications that allows for a low-restriction air cleaner and a cat-back exhaust (muffler). The bigger changes need more fuel to feed the extra power potential and that extra fuel and revised spark advance must come from reprogramming the computer or replacing the production

The DiabloSport device is called the Predator and is a works-with device, which has many tunes, or maps, pre-loaded in the device, including tuning for octane rating of the fuel. (Photo Courtesy DiabloSport)

DiabloSport offers many variations of its "inTune" kit for Chrysler products. They have many options and features. (Photo Courtesy DiabloSport)

Several years ago, programmable computers were considered to be racing units and AEM was one of the leaders in these advanced units. More recently these units have moved toward the hand-held units with more features. These units overlap street and race applications.

computer with an aftermarket high-performance computer.

MSD offers options for different applications. The MSD system is used by Indy Heads for its carburetor options. Manifolds are used for either throttle bodies (fuel injection) or carburetors.

Several MPI-style ECMs are available that work well for racing applications. One unique use is the Canadian Pinty's racing series (circle-track, road racing) because it requires an actual carburetor; no fuel injection.

With a carburetor on the fuel side, you need a distributor. Arrow Racing Engines offers a front cover that adapts a distributor (also available from Arrow). Fuel Air Spark Technology (FAST) has a Hemi system that splits the ECM into two

boxes: one for the fuel and one for the ignition. Therefore, if you just use the ignition box, you can still use the high-tech crank wheel as a built-in crank trigger system. Details are still being worked out.

In the MPI system the ECM controls everything; it is the heart of any engine performance package. There is no easy dividing line as to when you should change boxes or upgrade to racing hardware. Remember that the performance windows for any given package is only 5 to 10 percent wide and then the box must be reprogrammed or a new box added. In addition, remember that if you increase the engine's horsepower output by 100 hp or more, you probably need a new set of high-flow injectors.

Mopar Scat Pack Kits

Mopar Performance sells three high-performance kits for the various Gen III Hemi cars. These three kits offer the best example of reprogrammed performance. Each kit contains the exhaust, cam, and ported heads for significant horsepower gains. Each one must have the ECM reflashed before those gains are realized: 20, 56, and 75 hp. The reflash computer may only gain a few horsepower, but it opens the door for some very large gains when used with other performance parts.

Scat Pack 1

This kit includes a low-restriction cat-back dual-exhaust system plus a low-restriction air cleaner and a reflashed engine-control unit. This kit gains about 18 to 20 hp and 18 to 22 ft-lbs of torque.

Scat Pack 2

This performance kit is much bigger and requires the heads to be removed. The kit includes a high-performance cam, high-performance valvesprings, tie bars, required gaskets, and related hardware. The

AEM offers computers with a wide range of features. Typically, these ECM units or modules come with more hardware (shown), including wiring harness, protective shielding, and special connectors. (Photo Courtesy AEM)

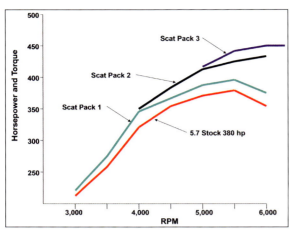

Reprogramming offers a wide range of features. The trick is to show the gain in terms of power that the owner finds useful. I thought that showing the Mopar Scat Pack packages was the perfect display device because they are three performance packages (hardware) that offer power increases and each one must be reprogrammed before the gain can be realized.

For the first few years of Gen III Hemi production, SCT reprogrammed the stock computer. Since then, SCT has a hand-held computer that talks to the car's computer through a USB interface cable. The popular unit for Hemis is

The baseline line was a 5.7 stock at 380 hp. The Scat Pack 1 has a cat-back exhaust and cold-air inlet system and gains about 20 hp. The Scat Pack 2 features a bigger cam and with kit number-1 gains a total of about 55 hp. Scat Pack 3 includes CNC-ported heads, high-flow exhaust manifolds and high-flow catalytic converters. This package added to the other two increased the power 75 hp over the baseline, but each kit (especially the last two) had to be reprogrammed before the gain was accomplished. This graph is based on Chrysler/Mopar test information.

called the SCT X4. It holds 10 custom tunes that increase horsepower and torque. These units can be installed by any of SCT's tuning dealers. The OEM computer stays in the car.

Aftermarket Options

performance gain is around 56 hp and 30 ft-lbs above the stock engine. The trick is that you must have the computer (ECU or ECM) reflashed and this is not included in the base kit. Arrow Racing Engines is one source that can do the reflash.

The reprogramming of these computers is constantly changing as far as what is offered and the modifications covered and power gained. Models and features are added on a regular basis, so check with the suppliers.

If you have a production vehicle and its production engine, the first performance step in the ignition area is to reprogram the stock computer. Scat Pack packages provide a good guideline, and reprogramming is required with each kit. If you get

Scat Pack 3

This is the most expensive kit. It includes a new set of CNC-ported aluminum heads, high-flow exhaust manifolds, and high-flow catalysts. With this kit added to the previous two kits, the total horsepower gain is about 75 hp and about 44 ft-lbs. This kit also requires a computer reflash and it is not emissions legal. Kits 2 and 3 both require the heads to be removed so you might want to plan ahead so that you only have to do it once.

Much of the performance testing and rating was done on a 425-hp base 6.1 package (420 ft-lbs) and it fell just short of the newer 6.4 engine that makes 485 hp and 475 ft-lbs.

FAST makes a lot of ignition kits and fuel injection hardware, including computers, distributors, and EFI systems. FAST is used by Indy Heads for its fuel injection version. I like the version that separates the fuel and ignition functions into two separate boxes because then you can use the ignition box with a carburetor to provide a very high-tech ignition. (Photo Courtesy Fuel Air Spark Technology)

AEM is a leader in programmable ECMs, which allow the owner to change the fuel and spark maps using a home computer. These basic units are very popular in racing. AEM now uses the "work-with" category of controllers, which work with the stock computer and do not replace it.

to the point where the horsepower of the engine has increased by 100 hp or more, you might consider upgrading to a new ECM. Today, aftermarket ignition manufacturers offer hand-held computers that work with the production onboard computer.

If you are past the low-restriction air cleaner/cold air system and the cat-back exhaust and have added some high-performance cam, spring, head, and header hardware, you could consider high-performance coils, but you must have eight. Also somewhere around the cam and head upgrades, you should consider installing colder spark plugs.

ECM

The first performance change after a few bolt-ons is to have the computer reflashed. SCT and Arrow Racing Engines offer this service. The work-with units offered by SCT, FAST, AEM, and others can also be considered. If you have several changes in mind spread over some time, the work-with units may be easier to use and less expensive in the long run. You should plan to have the computer reflashed after each major step or change. Major might be assumed to be horsepower gains in the 20- to 30-hp range.

A few years ago, programmable ECMs would have been an early add-on, but now with the reprogram options and the work-with units, these high-tech units are closer to racing and/or modified engines. I think that technology in these units was expanded to the work-with units for more user-friendly installations.

Coils

The production coils bolt to the valvecovers (105 in-lbs); one coil pack per cylinder. Each coil pack fires

Ignition System Upgrades			
The best ignition is still evolving. High-performance coils are just a beginning.			
Package	**Ignition**	**Modification**	**Upgrade**
5.7	Stock Computer	Reprogram*	High-Performance Coils** and High-Performance Plugs***
6.4	Stock Computer	Reprogram*	High-Performance Coils** and High-Performance Plugs***
5.7 and 6.4	Distributor/Carb	MSD or FAST	High-Performance Plugs***
5.7 and 6.4	Distributor/MPI	FAST	High-Performance Plugs***
* Reprogramming must be done at each power level, or step, equal to about 50 hp. ** Specific testing is incomplete at this time; high-performance coils from AEM and MSD. *** High-performance spark plugs are usually colder; the Bosch SPHR5, for example.			
Note that these ignition upgrades are not included in the "Performance Packages" in Chapter 6. Reprogramming must be considered part of every performance package.			

High-output, or high-performance, coils can be added at any time. However, the production coils are pretty good and the horsepower gains for the various high-output coils aren't clear.

two plugs and they fire the plugs at the same time. Each coil pack contains two coils.

Companies such as AEM and MSD offer high-energy bolt-on coils to replace the production units. AEM and MSD high-performance coils provide excellent performance and are two viable options. A lot of development is going on in this area, and I expect many new coils to be released shortly. Dyno testing and in-car evaluations are incomplete at this writing.

Distributor and Adapter

Many engine applications may require a distributor. Some racing

and competition sanctioning bodies require the use of a distributor or an owner may prefer to use a distributor. Because the MPI system controls the production Gen III engines, there is no distributor and no place to put a distributor.

The solution is a new front cover assembly from Arrow Racing Engines, which has a place for a distributor and a mechanical fuel pump. This front cover replaces the production front cover. The distributor is held at the front of the engine somewhat as on a Gen II Hemi. The front cover includes a special drive that attaches to the front of the cam and drives the gear on the bottom of the special distributor. You also need the distributor and the special plug wires that go with it.

Fuel and Ignition Control System

The high-performance 426 Drag Pak engine uses a FAST fuel and ignition control system. It is unique because the ignition part can be separated from the fuel section. If you just use this FAST ignition-only box with a carburetor, you would have a very high-tech ignition system even though you lost the fuel injection side. This approach

The MSD ignition and timing control box is designed to run the ignition on Gen III Hemis if they are converted to using a carburetor. It is recommended by Indy Heads to be used with the Indy Heads carbureted intake manifold. (Photo Courtesy MSD)

is being developed by Arrow Racing Engines for the Canadian Pinty's racing series for the 2017 race season.

System Upgrades

The problem with computers and electronics is that a 500-hp reprogrammed computer looks just like a 600-hp reprogrammed computer. A high-output coil looks like the stock coil except for the brand name. When distributors were used in racing, one of the ignition upgrades was a crank trigger. With all Gen III Hemis you do not need a crank trigger, you already have one, the wheel is inside the crankcase.

Arrow Racing Engines offers a special aluminum front cover that holds a distributor. It also holds a mechanical fuel pump. The kit comes with all the hardware that you need, including distributor drive gear, longer attaching bolt, plain timing wheel, cover bolts, and distributor clamp. In most cases, the sanctioning body rules that a carburetor must be used (mechanical fuel pump, middle left) and once the MPI sensors are removed, the ignition system needs a distributor (top left).

A crank trigger basically replaced the "when-to-fire" function of the distributor. In the electronic age, an eight-point star wheel inside the distributor that was about 1.5 inches in diameter controlled the when-to-fire function. The crank trigger wheel replaced this small wheel with a 6- to 7-inch-diameter wheel added to the front of the crank that had four blips on it (the crank rotates twice per cycle). On Gen III engines, you have a 32- or 58-tooth wheel that does basically the same thing; it is just that much more accurate. How to use this leading-edge technology best is still being evaluated.

Plug Wires

Spark plug wires are not used on Gen III Hemis except for the Indy Heads coils-under-intake conversion and the Arrow Racing Engines distributor kit. Contact these companies for the plug wire sets if required.

Spark Plugs

The spark plug's job is to ignite the fuel/air mixture and this becomes much more difficult as the engine's output and speed (RPM) increases. The Gen III engine's dual plugs are fired at the same time.

The new Hemis use a 1-inch-reach plug. Champion, Bosch, NGK, Denso, and Brisk make these plugs. Two styles of 1-inch-reach plugs are available: gasket seat and tapered seat. The 2009 and newer 5.7 uses the gasket-seat style of plug; the 6.1 and 6.4 engines use the tapered-seat style of plug. Older engines (2003–2008) also use the tapered-seat plugs.

For the average max-performance engine, I recommend the Bosch SPHR5 plug. If you use nitrous, you should use colder spark plugs compared to stock. The same holds true for supercharged and turbocharged engines: Use colder plugs.

When you remove the plugs from your engine, check to see if the plug has a gasket (gasket seat) or not (tapered seat) and be sure that you use the same style.

With some Indy Heads intake manifold offerings, the coils move under the intake manifold. They bolt to the bottom side of the intake. They sit above the sealed tappet chamber so everything is dry. With the coils relocated under the intake, spark plug wires must be used to get the spark back to the plug. The hardware for this unique conversion is available from Indy Heads. Once completed the engine looks like a dual-plug Gen II Hemi.

SUPERCHARGING

About five years ago, the next step in supercharger development was the entry of the OEMs. Ford, General Motors, and Chrysler released production versions for their high-performance cars. When Chrysler released the 6.2 Hellcat at 707 hp, the Dodge Charger and Challenger Hellcats became the fastest and most powerful American cars. In addition, they came with a full warranty, were emissions certified, and were very durable. These three OEM production versions have pushed the aftermarket manufacturers for more performance from their kits and changes have been happening quickly.

Although a supercharger delivers bolt-on performance, it does more than that. Street superchargers offer a flat torque curve, and that's desired in street engines. The install time for the kits is much less than other max-performance applications.

Years ago, supercharger manufacturers sold blowers, but now they sell kits that include everything that you need. These blower kits require that the engine's ECM be reprogrammed, but any number of ECM companies can do this. In some cases, the supercharger manufacturer includes a device with the kit, typically a "works-with" device. In addition, many ECM device manufacturers offer a "boosted" map for the fuel and ignition curves.

For example, (as discussed in Chapter 10), the changes for the Scat Pack 3 kit include heads, big valves, bigger cam, and other parts. It produces about 75 hp and 44 ft-lbs of torque. Typically, a street supercharger kit offers horsepower numbers around 100 to 120 and torque increases of 100 to 120 ft-lbs. That supercharger torque number is hard to match in any other way.

Knock Sensors

Each Gen III Hemi has two knock sensors, which communicate with the computer. If any detonation is detected, it pulls back the spark and saves the engine.

The Whipple 2.9 supercharger is required by the NHRA so that the Hellcat package can race in the NHRA Stock and Super Stock classes. The engine is also required to have 354 ci (5.8 versus production's 6.2 cc). The Whipple supercharger is also a twin-screw supercharger and slightly larger in displacement at 2.9 versus 2.38 cc. It looks similar to other blowers (front inlet, front belt drive), but it doesn't look as wide because it doesn't have intercoolers on each side. With the Whipple an engine's compression ratio is allowed up to 10.5:1. Although it could be driven on the street, this is mainly a drag race package and drag cars can use race fuel, which has a much higher octane rating and therefore can support the higher ratio.

Almost all engines with these street superchargers are basically new vehicles that include MPI. MPI engines tend to use the same group of sensors and have similar technology. The unknown factor is: Does this MPI technology allow these engines to have a supercharger added without concerns of the engine's compression ratio, octane requirement (pump premium is the general recommendation), and the related detonation issues?

Today's supercharger kits are designed to run on the street like the Hellcat. That means that the engine should run on pump gas, which is limited in octane, about 92 for pump premium. Detonation is a challenge for all engines, not just the supercharged ones. However, supercharged engines make more power and torque, which pushes the engine closer to detonation because they run the same fuel as naturally aspirated engines.

The engineers designing the Gen III engines added two knock sensors. If the knock sensors pick up any signs of detonation, they warn the computer and it lowers the spark advance so that the engine is not damaged.

Belt Drive

All supercharger kits, including the Hellcat, have a multi-rib belt that drives the supercharger. The Hellcat uses a 10- to 12-rib belt that's required to drive one of these 150-plus-hp blowers. If you have a five-, six-, or seven-rib belt similar to the production serpentine belt used to drive the accessories, you are going to have trouble.

The narrower belt will slip. On a dyno or at continued boost the slippage will melt and destroy the belt. On the street, with short bursts, the slippage drops the actual boost to about 4 psi instead of about 7 or 8 psi, which may be saving the engine from damage.

Supercharger Basics

If you are building or rebuilding an engine and plan to add a supercharger, I recommend that the engine be built for the specific supercharged application. Therefore, you would drop the compression ratio, so it's 8:1 or 9:1 and not 10:1 or 11:1. Use forged pistons if possible and plan to use premium fuel (92 octane). Also use premium rods, a forged crank, and premium piston pins. Copy the Hellcat: Use 9.5:1 max.

For performance and durability, you face two main issues with adapting superchargers to street engines: the engine's compression ratio and the blower's boost pressure. With a street engine, you will be running pump gas, which has an octane of 92. This limits the supercharged engine's compression ratio to about 9:1 max with MPI (9.5:1 on the Hellcat). Second, the supercharger's boost should be limited to a maximum of 6 or 8 psi. The Hellcat engine is an exception (see below).

The Hellcat front drive or blower drive uses its own belt, independent from the regular serpentine belt. The belt is much wider and has more grooves than the accessory-drive serpentine belt.

As a general rule, supercharger manufacturers try to match the supercharger size to the engine's size (cubic inches). You do not want the same-size supercharger on a 347-inch (5.7) engine and a 426-inch (7.0) engine. The 347 engine might use a smaller blower (about 2.3) and the 426-ci engines might want a larger blower (about 2.8, 2.9, or larger).

Most supercharger manufacturers offer many pulley sizes as a tuning aid for increasing boost pressure. Edelbrock offers six to eight different diameters. Caution: Most street blowers cause the engine to detonate

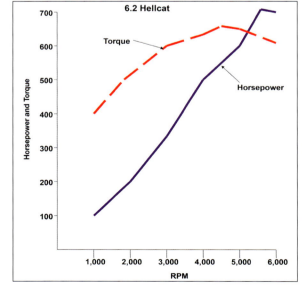

It is interesting to look at the Hellcat's power and torque curves. The 707 hp is amazing, but note that the Hellcat makes 411 ft-lbs of torque at 1,000 rpm, more than the 5.7 engine at its peak. The engine is fully emissions legal, with a full production warranty.

if the boost pressure is increased; 8 psi is a good maximum number. One solution is to use race gas, but it is expensive.

It is probably more important to install a 10- to 12-rib belt system to drive the supercharger than it is to change the pulley ratio if you have fewer than 10 or 12 ribs on the belt.

Intercoolers

Most of today's supercharger kits come with an intercooler, actually two, with one on each side. They are packaged with the supercharger in such a way that they are not visible once the kit is assembled. Basically, the intercoolers sit on top of the intake manifold with one 3 x 3–inch intercooler on one side of the supercharger housing and one on the other. Air goes through the blower, is compressed (increasing heat), passes through the intercoolers to take the heat out, moves into the manifold, and then into the head. The intercooler helps the engine run pump gas because it takes heat out of the inlet charge, which helps detonation. The Hellcat has an intercooler, as do the latest Whipple, Magnuson, and Edelbrock.

MPI System

There are a lot of different styles of blowers, especially street blowers. My main focus is superchargers designed for use with the MPI system because

Older 6-71 and 8-71 superchargers look like this with the pad on top that can mount two 4-barrel carbs inline. This style of blower was popular on the 426 Gen II engines. Indy Heads makes a blower intake manifold for use with this style of blower today.

all Gen III engines come with this system. This means that the throttle body mounts to the front or rear of the blower housing rather than on top of the housing. This gives them a much shorter, or lower, profile.

Aftermarket manufacturers make similar blowers for use with carburetors that mount on top. The whole area of blower design is changing fast as the hardware becomes more readily available and more manufacturers enter the market.

Carburetors

With a supercharger, dual 4-barrel carburetors are the most popular option. An 8-barrel system could be rated at 1,200, 1,500, or

This Edelbrock supercharger is shown in cut-away, so you can see the basically square intercoolers. One is placed on each side of the supercharger housing. This arrangement makes a neat and compact package.

1,700 cfm or more depending on the size of the two carburetors, 600 cfm, 750 cfm, 850 cfm, etc. Another issue with street superchargers is the length of the carburetor opening on top of the housing; it must be long enough for two carburetors inline. Manufacturers include Magnuson, Whipple Industries, and Edelbrock.

Edelbrock recommends using two 1405 Performer AFB carburetors (600 cfm, manual choke) for supercharged engines. Edelbrock also recommends the following calibration: primary jets, .101 inch; secondary jets, .101 inch; metering rods, .070 x .042 inch; step-up springs, 5 inches (orange); needle and seat assemblies, .110 inch. Edelbrock recommends this calibration for 347- to 426-ci engines.

Hellcat

The Hellcat engine was introduced in 2014. At 707 hp, it is currently the top dog in production horsepower in this country. Because it has been in demand and production has been somewhat limited, engines and production hardware have been difficult to find. So far, it has only been installed in the Challenger and Charger models, but that is rumored to be changing.

The Hellcat does not use the MDS that is popular in the 5.7 engines. The engine's displacement was reduced from 6.4 to 6.2 by changing to the 5.7's 3.58 inches (versus 3.72). Despite the displacement loss, the engine's torque increased from 470 to 650 ft-lbs in the Hellcat.

Chrysler says that 90 percent of the engine is new for the Hellcat engine, including the pistons, rods, crank, and piston pins. The cylinder block is the same as the 6.4 engine and the heads are basic 6.4 versions,

The IHI blower is a 2,380-cc reverse blower that has intercoolers. It regulates boost pressure to about 11.6 psi. This high boost pressure used with the engine's somewhat high compression ratio of 9.5:1 is accomplished by fine-tuning the MPI sensors, pulling back the spark in just the right places, adding a little fuel in the critical spots, having two knock sensors, having intercoolers, and having computer control of all the other loads on the engine. Without all that advanced technology and controls, the engine would be limited to 8 psi and a CR below 9:1.

but the intake valves (54.3 mm) are hollow-stem designs and the exhaust valves (42 mm) are sodium-cooled.

IHI manufactures the Hellcat supercharger and supplies the AMG division of Mercedes. At first glance this blower does not look unique from other street superchargers. The air enters the blower housing through the 92-mm throttle body mounted on the front of the housing. When it's in the blower, the air is compressed upward and then passes through runners where it makes an 180-degree turn and goes straight down on the outside of the housing through four heat exchangers (two for each cylinder bank, mounted inline) and then into the short, relatively straight intake ports.

A Few Supercharger Tips

The factory engineers have put in a lot of time figuring out which system is best for the street. One of the advantages of the twin-screw, or twin-rotor, supercharger is that it makes a flat torque curve, which is good for street driving. Another plus is that the typical 6- to 8-psi package doesn't run more engine speed (RPM), which is very good for street

driving. Most street superchargers peak at around 6,000 rpm, which is usable on the street. It is easy to use bigger parts to make the engine peak at higher RPM (7,500 or more), but these engines are more difficult to use on the street and generally require changes to the chassis, transmission, rear axle, converters, etc.

The Hellcat proves that 700 hp is possible on pump gas, premium, or 92 octane.

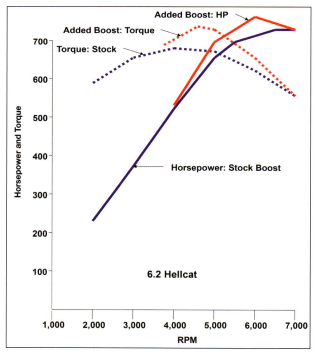

Engine Compartment

The supercharger sits on top of the engine so hood clearance is always a concern. In an engine swap situation, engine compartment size is also a concern because the supercharger and its hardware take up more space. However, with the front inlet and angled throttle body (used on the 6.4 and Hellcat), the extra height is limited.

Chassis

You should always consider the engine's effect on the chassis. The question to ask is: Can the chassis and tires handle another 120 ft-lbs of torque at launch? Basically, you will be going from 400 to more than 550 ft-lbs (the Hellcat is 650 ft-lbs). That is a giant gain. In the old days, this would be like taking out your small-block and installing a high-performance 440 or 426 Gen II.

Extra Boost

With any belt-driven supercharger, it is very easy to increase

With a supercharged engine, another possible upgrade is to increase the amount of boost pressure. For this test, I added 2 psi of boost pressure by changing the amount of overdrive on the pulley. There are several ways to change the pulley ratio, but I selected an ATI dampener with more overdrive. In this case, the engine added about 40 hp, but I strongly recommend better gas. I also think that any boost test should be run with headers and a cat-back exhaust on the engine.

the amount of boost for the 6- or 8-psi level. The real trick is to not be greedy. ATI makes dampeners that are the lower pulleys in larger diameters: in the 5- to 18-percent overdrive area. A larger crank pulley makes the belt travel faster and this increases the RPM of the blower.

Making the pulley on the blower smaller does basically the same thing: It makes the blower turn faster. Almost all supercharger manufacturers make smaller top pulley diameters for their blowers. For pulley selection for the production engine, Modern Muscle is a good source of information.

Remember that you need 10- to 12-rib belts to drive these 150-hp blowers and if you don't have one, changing pulley ratios (more overdrive) will only make it slip more.

Spark Plugs

Generally a supercharger needs to use colder spark plugs than does a stock naturally aspirated engine.

Aftermarket Sources

Selecting the correct aftermarket supercharger can be confusing because there are several sizes of supercharger, and one size does not fit all. Many manufacturers have offered a 2.3 unit and those have now been upgraded to a 2.9 unit. Some manufacturers already offer bigger ones and more are in development.

The Hellcat engine and the IHI supercharger package was the industry leader at its introduction in 2014. However, Magnuson, Whipple, Edel-

brock, and others have caught up. Developments will continue. The nice thing is that most of these supercharger companies offer complete kits for specific vehicles and engines.

Magnuson

The Magnuson blower is a twin-rotor supercharger with a liquid-to-air intercooler generating 6 psi of boost. Magnuson dealers and the Mopar Pro Shop offer these kits for Gen III Hemi models.

There are many kits for various models but I focus on the latest offerings for the 6.4 V-8 Charger/Challenger. (Magnuson also has kits for the 5.7 and 6.1 and for both cars and trucks). With the 6-psi boost, this kit shows gains of 120

Supercharger Upgrades

Every part of the engine, including the supercharger, can be upgraded. Here are the items that relate to the supercharger in general.

0. Use 10- to 12-rib belt drive
1. Smaller pulleys equal more boost; available from manufacturer or Modern Muscle
2. Bigger crank pulley equals more boost; available from ATI; if you change pulley size, the serpentine belt length may change
3. More boost means better fuel; 100+ octane
4. Port throttle body; more flow, less hassle
4. Bigger throttle body; more throttle bodies
5. Bigger volume supercharger; 2800/2900
6. Intercooler; almost all new blowers have an internal intercooler - add one if yours doesn't
7. Reprogram at each step
8. Bigger Injectors; may not be necessary if going from 400 to 500 rpm but required for 600 to 700 hp

ATI offers safety dampeners (SFI approved) for Gen III engines. In addition to the various standard engines, ATI offers overdrive dampeners for the Hellcat. They look the same except the OD is slightly larger. Note: The Hellcat (not shown) has two pulleys, one in front of the other. The wide pulley is for the supercharger. (Photo Courtesy ATI)

The stock Hellcat engine seems to have a very complicated front drive system when you first look at it. The supercharger has a separate drive belt from the rest of the accessories that are driven by the standard serpentine belt. The supercharger belt is wider and its pulley is at the top. The next two pulleys (top left) are also part of this system. The crank pulley is behind the front serpentine pulley and it is somewhat larger in diameter but difficult to see.

The latest Magnuson offering has dual single inlets and dual throttle bodies. The supercharger is a larger unit and became available in early 2017.

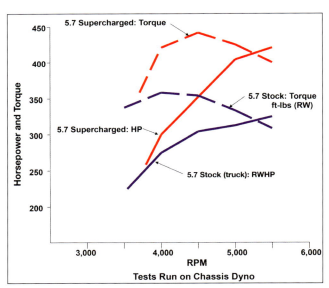

The early 5.7 Gen III Hemi is the lowest horsepower and lowest torque of all the Gen III engines and therefore in need of a supercharger more than any other engine. Using a chassis dyno (horsepower numbers are conservative relative to an engine dyno), the baseline 5.7 was stock and made about 325 hp. When the Edelbrock supercharger was added, the power increased about 100 hp and the torque increased about 80 ft-lbs. The stock 5.7 had 323 hp and the torque was 362 ft-lbs.

hp and 120 ft-lbs of torque using a liquid-to-air intercooler. This kit comes with a DiabloSport Trinity handheld programmer (basically a works-with unit). This kit was tested on a chassis dyno at 550 rwhp and more than 500 rwtq.

Although this unit is a 2300 supercharger, Magnuson showed a 2.9 or 2900 unit at the 2016 SEMA show (to be available for 2017) along with an even bigger twin-snorkel, twin throttle body unit for mid-2017. One interesting aspect is that the Mopar Pro Shops shows that some Magnuson supercharger kits may carry a CARB number along with being 50-state emissions legal for on-road use.

Edelbrock

The Edelbrock supercharger is a twin-rotor unit with four-lobe rotors with a 160-degree twist. Edelbrock currently offers supercharger kits for Ram trucks and Chrysler 300, Challenger, Charger, and Magnum cars. These superchargers come with an air-to-water intercooler. The boost pressure is around 5 to 6 psi.

Several levels are available, including: Stage I for street systems and Stage 3 for Professional Tuner systems. The Stage I tune works well for 91-octane pump premium gas. Edelbrock supercharger kits come with a handheld tuner (works-with

style). These kits are 50-state legal and have a CARB number when used with the correct Edelbrock tune.

Also available from Edelbrock is the 2650 rotating assembly, which is 15 percent larger than the current 2300.

The Edelbrock E-Force supercharger is packaged similarly to the other versions with the intercoolers on each side of the blower housing. It has the drive pulley at the front (right side center) and the angled-to-left throttle body. (Photo Courtesy Edelbrock)

The Magnuson supercharger package for Gen III Hemis is similar to the IHI and the Whipple. The company offers several sizes that come with the intercoolers packaged on the sides of the blower. Magnuson also offers a larger blower than the IHI (close to the Whipple) that looks similar to the IHI with its angled throttle body.

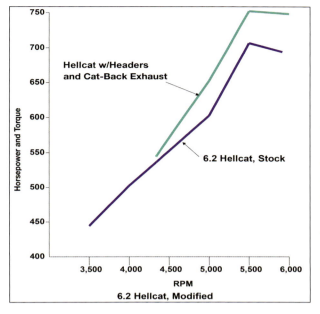

Perhaps the easiest upgrade for a super-charger engine is to add headers and a cat-back exhaust. Using the Hellcat 6.2 as the baseline (707 hp) the engine picks up about 38 hp for adding American Racing Headers cat-back exhaust and 1⁷⁄₈-inch headers.

The ProCharger kit by Modern Muscle is some-what unique even though it is belt driven. The ProCharger is not a gear or rotor supercharger but a vane supercharger or centrifugal unit.

Supercharger Packages

This whole group is in the early development stages and the hardware is still evolving.

Superchargers were not included in "Performance Packages" in Chapter 6 because they are changing so quickly.

Package	Number 1 Upgrade	Number 2 Upgrade	Number 3 Upgrade
Hellcat	Ported Throttle Body	Headers and Cat-Back	Bigger Blower
5.7 SC*	Blower Package	Bigger Throttle Body (ported)	Headers and Cat-Back
6.4 SC*	Blower Package	Bigger Throttle Body (92 to 95 mm)	Headers and Cat-Back
426 SC*	2.9 Blower	92- to 95-mm ported Throttle Body	Headers and Cat-Back

* The first choice for a blower package is the production Hellcat, but it is not currently available in package form; perhaps it will be available soon from Mopar Pro Shops.

Whipple

The Whipple supercharger is a twin-screw design and looks similar to other street blowers; front inlet and single throttle body. Whipple offers kits for Gen III Hemi cars, Ram trucks, and Jeep Grand Cherokees. Many performance levels are available from blower only to those with an intercooler. The Whipple is also used on Drag Pack race cars. Whipple will be offering a 4.5 blower shortly.

Kenne Bell

The Kenne Bell supercharger is a twin-screw design and the current 6.4 Hemi kit uses a 2.8 or 2800 size blower with an intercooler. The company has several other sizes, including a 3.6 and a 4.2. The unique feature of the Kenne Bell supercharger is that it is not a front-entry system. With the Kenne Bell supercharger the inlet air comes in the rear of the supercharger.

Typically a 180-degree elbow called the mammoth inlet (an aluminum casting) is used to turn the air around the corner and into the rear of the housing. This elbow is twice as large as one 80-mm throttle body and apparently flows twice as much air. This allows placement of the actual air inlet in cold air rather than using underhood air. They use 6-psi pressure on the 6.4 and 7 psi on the 5.7 and 6.1 engines because of the compression ratio differences (10.7:1 on the 6.4 and 10.2:1 on the 5.7). This kit makes about 132 hp more than the stock engine on 91-octane gas at 6.5 psi.

Modern Muscle

The ProCharger supercharger from Modern Muscle is somewhat unique among street superchargers, including the production Hellcat. The ProCharger is a centrifugal supercharger rather than a twin-screw or twin-rotor design. A belt off the nose of the crank drives it, similar to the twin-screw/rotor units. It has a gear set (4.1:1) that gets the basic speed up to the high numbers for maximum boost and power.

Modern Muscle offers kits for the 5.7, 6.1, and 6.4 Hemi engines in all the cars and the Grand Cherokee Jeeps. Most street kits use 7-psi boost. Intercoolers and a tuner kit are offered as options. Horsepower gains are in the 160 to 200 range depending on model. The boost is through the front throttle body and a stock intake manifold is used. Most of the kits are 50-state emissions legal.

Vortech

The Vortech supercharger has a self-contained lubrication system. It is similar to the ProCharger because it is also a centrifugal supercharger. The Vortech unit uses a 3.6:1 step-up ratio along with helical-cut gears for quieter operation. This unit has a 3.5-inch inlet OD, a 2.75-inch discharge OD, and a 2.38-inch discharge ID.

Turbochargers

Turbochargers are similar to superchargers except that exhaust gases drive the exhaust turbine and this drives the intake turbine rather than a belt from the crank. Garrett and Turbonetics, for example, offer turbochargers, but remember that they sell bare turbos rather than complete kits as supercharger manufacturers currently do.

Nitrous Oxide

Nitrous oxide kits are popular and reasonably easy to install. Nitrous companies offer kits up to 500 hp. Street kits are close to 100 hp: remember that you must inject more fuel along with the nitrous. Some "work-with" handheld computers have a "use-with-nitrous" map.

This turbocharger is from Garrett. Although supercharger suppliers provide complete kits, turbocharger suppliers supply only the turbocharger.

This is a Vortech supercharger on a Jeep Grand Cherokee SRT8. This package should be very popular because the Grand Cherokee has AWD to get all the power to the ground. With this supercharger on the SRT8, lots of power will be available. (Photo Courtesy Vortech)

A nitrous kit for MPI engines is based on the single, large, round throttle body and can use this style of kit. The plate installs between the throttle body and the intake and it includes two solenoids.

EXHAUST SYSTEM

Because all Gen III Hemis have been built in the past 12 to 14 years, they all use a catalytic converter and generally have a low noise level. However, although it isn't noticeable and you can't hear it, a lot of technology went into the current exhaust systems.

The only part of the exhaust system that physically touches the engine is the exhaust manifold, but you must look at the complete system to properly evaluate its effect on the engine. A lot of the rest of the exhaust system past the exhaust manifold, such as pipe length and diameter, the shapes and bends and interconnections, and exhaust tips, all can affect the engine's performance, torque, and horsepower along with the sound.

Today's performance exhaust system must include a high-flow catalytic converter, one in each pipe if it is a dual-exhaust system, which is common on passenger cars. The typical production exhaust manifold is quieter than tubular headers that are often used to replace production manifolds.

The 426 Gen II Hemi engines did not use a catalytic converter; that came a few years later. The catalytic converter is usually located near the exhaust manifold outlet and a little farther back, next to the transmission. Today we have cat-back exhausts, which are 50-state legal and can replace the production exhaust system rearward of the catalytic converter. This little change allowed the technology in the exhaust system to

jump ahead and brought out much new hardware in the exhaust system but not as much in the manifold itself. Other aspects brought changes in that area.

Mopar Performance offers this high-flow exhaust manifold as part of the accessories for its crate engines.

This special exhaust manifold cut-away was made for display. The outer shell has been cut away so you can see the individual tubes that make up the actual exhaust manifold. It looks an awful lot like a shorty header.

The standard Gen III exhaust manifold looks smooth but is not a casting. It is a jacketed construction (one inside the other), but it is not easy to see.

Manifold

The exhaust manifold bolts to the head (23 ft-lbs torque) and that is the only part of the exhaust system that physically touches the engine. The typical Gen III Hemi exhaust manifold is jacketed, which means it is double-walled. It is fairly short. The only upgrade seems to be the Jeep/SRT version (77072462), which is a standard casting.

Catalytic Converter

All production exhaust systems include a catalytic converter. Flowmaster makes high-flow converters. Several exhaust system manufacturers, such as TTi, Kook's, and American Racing Headers, offer exhaust systems with the converter included.

The catalytic converter, usually one per side, is about the size of a coffee can and sits immediately at the end of the exhaust manifold, next to the transmission. The spark-plug-looking device in the center of the converter is the oxygen sensor, which talks to the computer. If you install headers, you must transfer this sensor to the header.

Exhaust Bolt Pattern

Gen III Hemis use different left and right cylinder heads. One of the differences is the exhaust bolt pattern. To envision the difference, picture a 3-inch-square box with an attaching bolt hole at each corner. There are four individual ports so there are four boxes on the exhaust face. For the center two boxes and the one to the left, on the driver-side head, the upper left bolt and the lower right bolt are threaded; the other two are not used. On the passenger-side head, the center two boxes (ports) and the one to the right, the upper right bolt and the lower left bolt are threaded and the other two are not used.

Exhaust manifolds and headers are unique to each side to fit into the engine compartment and can't be switched. If the heads are reversed, you have a problem.

The left and right heads are not the same. The exhaust flange is not machined the same; the bolts around the center two ports are unique (see Chapter 7 for more information). This is the front view of the heads on the engine. The head on the left is the passenger-side) head; it has two machined holes, one large core plug, and a notch or hole where the rib is. The driver-side head has three machined holes to mount the power steering.

This is the rear view of the heads. Here, the driver-side head is on the left. It has the same three holes in the pad plus the notch for the rib. The passenger-side head has the same pad (power steering mount) but no drilled/tapped holes.

Engine Swaps

Gen III Hemi engines are popular for engine swaps, especially in muscle cars. Several header manufacturers offer headers and exhaust systems for these swaps. The production jacketed exhaust manifolds can be used, but clearance issues may have to be resolved.

Headers

A header has four individual tubes that are tuned to the specific performance requirements of the engine and the intended application (these replace the exhaust manifold). Each tube is about 1.5 to 2 inches in diameter and generally more than 24 inches long.

Performance gains for tubular headers and low-restriction exhaust systems vary mainly because stock exhaust systems used on production

Calculate Header Cross-Section

An equation can be used to calculate the diameter (size) of the primary tubes in the engine's headers. It is based on the volume of one cylinder and the RPM at which peak torque occurs in the engine.

$$PA = (CV \times RPM) \div 88,200$$

or

$$RPM = (PA \times 88,200) \div CV$$

Where:
PA = primary pipe's area (square inches)
CV = the volume of a single cylinder (cubic inches)
RPM = the engine's RPM at peak torque

Once you have calculated the primary pipe's cross-sectional area, you need to convert it into a pipe size. You can use the following formula.

$$PA = D \times D \times .7854$$

Where:
PA = cross-sectional area
D = ID of the header pipe
.7854 = math constant
1.273 = math constant conversion factor? (inverse of .7854)
Header Pipe = OD = ID + 2 x wall thickness)

In most dual-purpose engines, the power peak is about 1,500 to 1,750 rpm above the torque peak.

For example, a 5.7 engine might make peak power at 5,750 rpm. The peak torque can be estimated at 4,000 (going from 5,750 to 1,750). One cylinder's volume is 43.37 (347 ÷ 8). Therefore the cross-sectional area of the pipe is 1.97 square inches (43.37 x 4,000 = 173,480; then 173,480 ÷ 88,200 = 1.97). The actual header pipe size is 1.58 inches (1.97 x 1.273) x .5). The typical header tube wall thickness is in the .040- to .060-inch area, so the actual outside diameter of the header pipe is 1.66 inches (1.58 + .040 + .040).

Performance Trends offers computer programs that can calculate all of this with more accuracy and much easier than using a calculator and formulas. Unfortunately, header manufacturers offer only certain header sizes, in both primary length and tube diameters.

Therefore, in the above example in which you calculated that the header size should be 1.66 inches, the closest size is 1⅝ inches. The smallest header offered today is 1¾ inches.

The other trick that is somewhat misleading about these calculations is that installing a 2-inch header on the above engine package is not going to make this 5.7 engine package peak at 6,500 (5,000 peak torque).

Kook's Headers has a chart for header sizes based on engine horsepower. It shows a 400- to 550-hp engine with 1¾-inch headers, a 550- to 750-hp engine with 1⅞-inch headers, and a 750- to 850-hp engine with 2-inch headers. The stock 5.7 engine is at 385 to 390 hp (pretty close to 400), the stock 6.4/392 is at 485 (pretty close to 550 with a mod or two), and the 426 (crate engine) is at more than 750. ∎

vehicles vary greatly because of body shapes, lengths, and trucks versus cars.

Another related aspect is that Gen III engines make a lot of power without having much larger displacement. Around 5 percent is a good baseline horsepower gain. The typical 5.7 has around 400 hp and the 6.4 has around 485; that puts the power gains in the 20 to 25 hp, which is about right. The problem is that most street-header tests are run with a full exhaust system and the header and exhaust system seems to gain about 6 percent, which is 24 to 30 hp, which is what is found for cars. Exhaust tuning can trick you, so it could be 50-50 for the header versus exhaust pipes (3 percent each).

Caution: On Gen III Hemi engines because they are MPI and computer-controlled, performance changes (headers and a cat-back exhaust, for example) will put the engine on the edge of being too lean. If it is too lean, it could lose performance rather than gain it.

Generally, the stock ECM can handle the addition of headers, but if other engine upgrades are performed, headers can push it over the edge. Solution? Reprogram the ECM. If the exhaust system is only one of several modifications planned, also plan on reprogramming the computer.

Several types of tubular headers are available: 4-into-1, Tri-Y, Step, and Shorty. Technically, the Step and

Shorty headers are special versions of the 4-into-1. They are typically made with a flange bolted to the head and four steel tubes welded to the flange. The "how long" and "how many" vary and they join together in the collector.

4-into-1

These are the most common headers and have four individual tubes per side that carry exhaust gases from each exhaust port in the cylinder head to the collector. These four tubes are called primaries. The tubes vary in length from 24 to 36 inches, depending on design. TTi, Kooks, American Racing Headers (ARH), and Hedman all offer top performing exhaust systems for the Gen III Hemi.

Gen III street versions, which are somewhat easier to install, have a smaller primary diameter (around 1¾ inches) and a moderate diameter on the collector (around 2½ or 3 inches). Gen III racing versions usually have larger primary diameters (around 2 inches or bigger), but race packages are still being developed. With a 4-into-1 header, you want the pipes to be equal in length, which is difficult to do on a production car.

Making the primary tubes equal in length and fitting them into an engine compartment is even more difficult. The cylinder tube closest to the radiator tends to be too long and the cylinder tube closest to the firewall tends to be too short. For street headers, the compromise is toward less equal length and easier fit. Race headers go the other way.

Tri-Y

A Tri-Y header design starts at the head with four tubes matched to the four exhaust ports at the flange similar to a 4-into-1. After the four tubes come out of the flange, they merge into two tubes and then the two tubes are joined at the collector. This creates a 4-into-2-into-1 design that has three steps. The various joints look like "Ys," which is where the header got its name).

This design was generally used for street headers because it was easier to fit into the engine compartments. In the 1970s and 1980s, they disappeared from the marketplace.

In the mid-1990s, SEMA asked the California Air Resources Board (CARB) to define a test program for performance parts so they could be sold for street use. Each part would be tested and if approved, assigned an emissions exemption number. Header manufacturers jumped at the chance to have street-legal exhaust products.

The manufacturers looked at various designs on the dyno focusing on street vehicles and street use rather than race cars.

They found that the Tri-Y design offered more torque and they could match or improve on the power of the 4-into-1 design. This performance gain on both ends of the performance curve was accomplished by adjusting the tube diameters and their respective lengths.

Shorty

The shorty header is technically a 4-into-1 design, but instead of having 30-inch length primaries, it has 6-inch primaries. It is sometimes called a block hugger. The shorty was originally made for street rods because you usually install big V-8 engines into small engine compartments that were designed for 4s and 6s.

The biggest advantage of the shorty design is the ease of installation

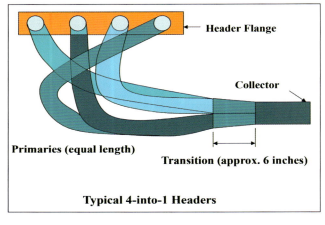

Typical 4-into-1 Headers

Header Flange

Collector

Primaries (equal length)

Transition (approx. 6 inches)

Typical Tri-Y Headers

Header Flange

First Section (4 pipes)

Third Section (1 pipe)

Second Section (2 pipes)

The 4-into-1 header is the industry standard. Getting the four tubes to be somewhat equal in length and also fit into the engine compartment makes it complicated.

Tri-Y header installs are easier, but they are difficult to find as non-custom parts.

This is a shorty header made by BBK. This style of header is popular in street rod and engine swaps because it takes up less space and is easier to install. (Photo Courtesy BBK)

This is the most complicated shorty header that I came across and shows how much work it is to install equal-length tubes into an engine compartment. (Photo Courtesy BBK)

Hedman offers this header, called a long tube that looks between shorty and long traditional hardware. I would call it an intermediate. It should be easier to install than the traditional long-tube. (Photo Courtesy Hedman Hedders)

American Racing Headers offers this long tube or 4-into-1 header for the Hellcat (left side). (Photo Courtesy American Racing Headers)

These are ARH Hellcat headers for the passenger's side. Although the headers can't be reversed, there would be problems if the heads were reversed. (Photo Courtesy American Racing Headers)

Kook's offers a set of stainless steel headers for the newer 5.7 in the Ram 1500. (Photo Courtesy Kooks)

and fit into the engine compartment. In many cases a shorty header looks very similar to a production exhaust manifold except that with the shorty header you can see four individual pipes whereas the exhaust manifold merges the tubes. Short headers do not make as much power as the other two types of headers.

Step

Technically, step headers are another version of the 4-into-1 header. With a typical race header with all primaries designed at the same length, the tube size is constant, at 1¾, 1⅞ inches, etc. The 4-into-1 primaries are larger and the length changes tuning for different performance characteristics (peak torque, peak RPM, etc.), but the tube diameter stays the same for a given design. With a step header the primary tube changes size after 6 to 12 inches (this length is variable) and may change size again before it gets to the collector (usually just one step is used).

The size of each is these steps (diameters) and the length of the tube in each step is determined by extensive testing on an engine dynamometer. Each set of sizes and lengths is unique to the specific engine and RPM package.

The manufacturer's reason for all this testing is to gain the best of both worlds: best torque while maintaining the best power. For example, TTi currently offers a step header for most Gen III cars with a 1⅞-inch step going to a 2-inch step, which is

Exhaust System Upgrades

Headers are the upgrade, but there are several sizes so each package has a preferred size that works best with that cylinder head.

Package	Best Head	Header	Exhaust	Exhaust Pipes
5.7: 1, 2, 3, 4	Eagle and Ported	1¾ inches	Cat-Back	2.5 or Bigger
6.1 and 6.4: 5, 6, 7, 8	Apache and Ported	1⅞ inches	Cat-Back	2.75 or Bigger
Optional for all engines	All	Step	Cat-Back	2.75 or Bigger
426 or Larger	Ported Big Valve	1⅞, 2 inches	Cat-Back	3.00
426+ Optional	Ported Big Valve	Step	Cat-Back	3.00

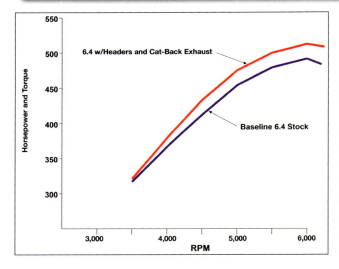

For a header test, a 6.4 engine was selected; the baseline was the stock exhaust system. When ARH headers and a cat-back exhaust were added, the power increased by about 24 hp, which put the 6.4 engine over 500 hp in a very mild trim.

described as a long tube (about 30 inches) and it has been tested to gain 26 hp.

Collectors

The collector is the end of the header and is a straight piece of tubing about 2½ to 3½ inches in diameter. In a 4-into-1 header, the four pipes come together in four round circles in a rounded square and become a large round circle. Generally, large tube headers receive larger collectors, and thus 1¾-inch headers use a 2¾-inch collector and 1⅞-inch headers use a 3-inch collector. The collectors are basically the same for 4-into-1 and Tri-Y header designs.

The typical exhaust header collector is a straight and round tube about 2¾ to 3 inches in diameter and about 12 to 24 inches long. A merge collector is tapered on both ends: forward toward the tubes and

A merge collector is basically tapered on both ends. Although they can be incorporated into almost any header, they are more expensive to construct so they tend to be used mainly in racing.

toward the rear. When the four primary tubes come together, the center point of the tubes is extended (about an inch or two) and tapered to a point. The outer wall of the collector tapers slightly over a short distance to the center of the collector. Then the collector expands until slightly larger than the standard collector.

The length of the taper is changed for different tuning affects, but the short ones are about 6 to 8 inches long and the long ones are about 18 inches long.

High-flow and high-velocity exhaust gases in the merge collector increase the amount of torque and broaden the torque curve. However, they are more difficult to fabricate and therefore more expensive. Many header manufacturers offer a merge collector on some models but not all because of that extra expense.

Oxygen Sensor

Oxygen sensors are installed in all MPI engines including Gen III engines since 2003. The oxygen sensor measures how much oxygen is in the exhaust pipe. This oxygen output allows the ECM to function properly. The ECM controls both the amount of fuel and the amount of spark that the engine receives. In some cases (including production exhaust systems) there may be more than one oxygen sensor in the exhaust system.

If any style of header is installed on a vehicle, the oxygen sensor must be reinstalled in the new headers. The manufacturer should have added an oxygen sensor fitting, typically in the collector. If the sensor must be moved, both TTi and Modern Muscle offer extenders for the wire. If you are building a MPI engine, you want the oxygen sensor in the header.

Coatings

There are a lot of different types of coatings for headers. Although it may look like paint, it is not simply spray-painted. Some coatings are for appearance (such as paint), and some are a form of heat shield or thermal barrier. Coatings can be ceramic, polished ceramic, or nickel-chrome plate.

Although some companies specialize in various coatings, most exhaust coatings are offered by the header or exhaust manufacturer. Header manufacturers also offer thermal barriers that are applied inside the tubes.

Although each type of coating offers unique features, the main focus of a heat barrier (outside or inside) is to keep the exhaust gas heat inside the tube and not allow it to heat up the engine compartment.

Mufflers

Before the catalytic converter, a muffler was the largest source of backpressure in the exhaust system for performance engines. Standard engines had single exhausts and small exhaust pipe diameters,

The trick with a muffler is to keep the backpressure low and still lower the noise level. This short muffler from Flowmaster helps the muffler fit under the car or truck. Flowmaster offers entrances in several locations; this one is centered on both ends.

which added lots of backpressure. Performance engines received dual exhaust pipes. In the muscle car era, high-performance engines wanted to lower exhaust backpressure so headers were added and then they went after the mufflers.

The early answers were the Street Hemi muffler (1966 through 1971) and the Imperial muffler (from about 1968 to 1970). The Imperial was good for performance, but it was larger (longer mainly) than the Street Hemi and was hard to fit into smaller vehicles, such as A- and E-Body cars, even many B-Body cars.

The turbo Corvair muffler was simply called the turbo muffler. It was smaller and easier to fit under cars and had very low backpressure.

When big engines went away in production (1972), the demand for trick, high-flow mufflers stopped. The real turbo muffler became difficult to find so most exhaust manufacturers made their own. Many racetracks started requiring race cars to have mufflers in both drag racing and circle-track racing. That requirement encouraged manufacturers to look for small, lightweight mufflers that didn't hurt engine performance. An empty box with a hole in each end still had backpressure and was noisy.

Then manufacturers began using sound wave science. The real push came when the CARB defined the cat-back system and that it was legal on street cars. This occurred in the mid-1990s along with emissions headers and the market exploded.

Manufacturers began designing and developing low-restriction, lightweight, small, and easy-to-install mufflers that actually lower the engine's backpressure and still kept the noise at reason-

able levels for real street use. These are available today from TTi, Borla, Kooks, ARH, Flowmaster, Corsa, and Hooker.

Exhaust Pipes

In general the exhaust pipe includes everything rearward of the catalytic converter and muffler. In some cases the muffler is near the rear axle assembly and the exhaust pipes are rearward of the exhaust manifold and converter. In the old days, big-blocks received big exhaust systems and the 440 and 426 Gen II engines received 2¼- or 2½-inch exhaust pipes.

Gen III exhaust pipes are 2.75 inches (Hellcat) and high-performance cat-back systems are 2.75 or 3 inches made by Kooks, TTi, ARH, Borla, and Corsa. In many cases, these high-performance exhaust systems come with converters, mufflers, and pipes.

H-Pipe

The H-pipe is a short tube that is used to connect the driver- and passenger-side exhaust pipes. The original tuning science was to place it where you wanted the header's collector to end. Therefore, if dyno testing showed that a 12-inch collector was the best package, you placed the H-pipe or cross-over pipe at the end of the 12-inch collector. In reality the H-pipe typically crossed under the transmission crossmember or just to the rear of that, which put it under the transmission output shaft and housing. With the typical header installed on a car, the typical collector ended near the transmission crossmember.

H-Pipes are available from TTi.

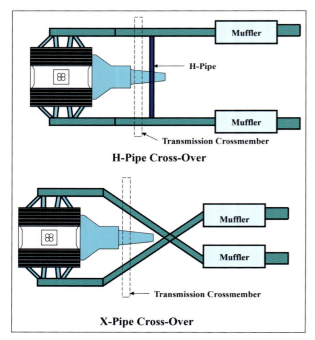

H-Pipe Cross-Over

X-Pipe Cross-Over

The exhaust pipe is supposed to get the exhaust gases out the back of the vehicle. The basic H-pipe connects the left and right side to help with engine tuning, which was introduced on the high-performance muscle cars of the 1960s. The X-pipe has only been around the last few years. You would only use one.

X-Pipe Cross-Over

The X-pipe cross-over is a recent development, which has its roots in racing. The X-pipe is another form of H-pipe interconnection; another way of connecting the driver- and passenger-side exhaust pipes for tuning purposes.

With the H-pipe, the two main pipes are straight and the crossover is straight, forming the H. With the X-pipe, the two main exhaust pipes are curved toward the center of the vehicle until they actually touch and then curve away from each other back to the original position, forming a large X. The center of the X (where they touch) is cut open on both pipes. Then, the two pipes are welded together sealing the two holes, but keeping the pipes in contact.

X-pipes are also made in 3- to 4-foot sections all welded together that can be installed in almost any exhaust system. In general, manufacturers make X-pipe kits in 2¾- and 3-inch pipe sizes so that it can match up to the exhaust system.

To date, dyno tests are inconclusive; the X connection is usually farther rearward on the car than the H and I think the extra distance helps the engine's tuning. Most tests are performed on chassis dynos rather than engine dynos.

Dynatech, Corsa, TTi, Kooks, ARH, and Hedman, among others offer X-pipes and/or cat-back systems that have an X-pipe.

Exhaust Gaskets

Sealing the exhaust manifold to the cylinder head can be a challenge. Production manifolds are generally easy to seal. Manufacturers have designed headers for use on street cars and have made the headers stiffer, especially the flanges. They use thicker, high-temperature exhaust gaskets. The high-temperature material is stronger and more resistant to leaks and the extra thickness allows for more flexibility. On Gen III Hemis, a bolt attaches to each side of each individual port.

In my opinion, having the exhaust ports separated, as with the Hemi design, allows for better sealing of the exhaust manifolds. However, the exhaust sealing problems on wedge engines almost always occurred in the middle where the two exhaust ports sit right next to each other; this doesn't occur on the Hemi.

Cat-Back System

Cat-back is a term that came about in the late 1990s to refer to exhaust systems that changed the items rearward of the production catalytic converter: mufflers, exhaust pipes, and interconnection.

Borla, Flowmaster, TTi, Kooks, Corsa, Hedman, ARH, and Kooks offer systems for the New Hemi, and many models are offered for the Charger and Challenger. However, all manufacturers don't offer systems for the Grand Cherokee.

Basically, on vehicles built after 1995, a cat-back system reduces the overall exhaust system's backpressure by 50 percent. This change can offer potential gains in the 5- to 7-percent range, which would be 24 hp (the average is 6 percent) on a 400-hp 5.7, which chassis dyno tests confirm. In some cases, the power output may be similar, but the exhaust note can be very different and that comes down to personal preference.

When working with cat-back exhaust systems, stick with one manufacturer for the complete system. Dyno tests that show 15- to 20-hp gains are performed on complete systems by one manufacturer. That allows the tech to tune those parts to give the engine added performance, which might not occur on a mix of parts from several manufacturers.

ENGINE BREAK-IN AND TIMING

When your Gen III Hemi is assembled and ready to run, the engine needs to be broken-in before it is put into service on the street or track. You can do that on a dyno, in-the-car, or on a hot stand. Dynos are very common today and just about every engine shop has one. The hot-stand just runs the engine in and doesn't measure loads, just temperatures and oil pressure as a car normally does. I will assume that you plan to use a dyno for the engine break-in.

Dyno Overview

The dyno is a tool used to measure engine performance. There are several types of dynos, including chassis dynos and engine dynos. The chassis dyno measures performance at the rear wheels and tests the engine as installed in the vehicle. An engine dyno tests the engine by itself in a dyno cell (special room). Dynos themselves haven't changed much in the past 50 years, but the instrumentation that controls it is completely different and constantly changing.

Because they are so readily available, using a dyno for your engine break-in cycle is realistic. Today's dynos have engine break-in cycles programmed into the basic software.

Dyno engine mounts are adjustable so it can be used with almost any engine. The other aspect of adapting your Gen III engine to a dyno is the bellhousing that mates the engine to the dyno. A small plus here is that the Gen III Hemi bellhousing bolt pattern is the same as all the Mopar small-blocks, both the 1964–1993 A-engine and the 1992–2003

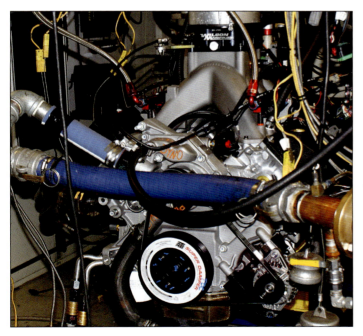

With all of the sensors used in today's engines and having to know information about all eight cylinders at the same time makes the dyno cell look like the inside of a computer.

A dyno operator sits outside the actual dyno cell behind computer screens along with the various controls.

Magnums. Gen III engines and race versions use an eight-bolt crank; most small-blocks use a six-bolt.

The Amsoil Engine Masters Challenge is an all-makes engine dynamometer competition. This 400-inch Gen III Hemi was built for the 2017 contest, and it performed well in later competitions.

Your first step should be to run all of the in-car systems: exhaust manifolds or headers, fuel system, oil pan and oiling system, ignition/computer system, and water pump. On MPI Gen III engines the ECM is important, along with all of the sensors. Focus your dyno time on the engine break-in cycle then a quick power check, and finally install the engine in the car.

During the break-in cycle, always check for fluid leaks, including gas, water, and oil. Gas is very important with MPI because you have many junctions to check. Water leaks are most common from the front of the engine and the area around the water pump and front cover. Oil can leak from the oil pan (all four sides) and the valvecovers. There is no oil in the intake manifold area, but it can leak air and that can cause problems.

After the engine is bolted to the dyno, hooked up to the proper hardware and if all of the electrics are plugged in (temperature, oil pressure, MPI sensors, etc.), the engine is ready for its first test. Be sure to check the oil pan to verify that it has enough oil. Use a conventional oil such as from Valvoline or Pennzoil or a break-in oil such as from Comp Cams with an oil weight of 10/30 or 15/40. Change this oil after 500 miles. The check-the-pan test is even more important if you plan to use the in-car engine break-in approach.

Next, fill the cooling system and be sure that there is no air in the system. Before you start the engine, prime the oiling system by rotating the engine clockwise (by hand or by disconnecting the coils). Oil should get up to the valvetrain before starting.

Compression Test

To run the test, follow this procedure:

1. Run the engine up to normal operating temperature (180 degrees F) and then shut off the engine
2. Pull the center coil wire (so the engine cannot start accidentally)
3. Remove the air cleaner
4. Remove all eight spark plugs
5. Open throttle blades and hold open (1/2 throttle to wide open) to allow air to enter
6. Thread the compression tester into the number-1 spark plug hole
7. Crank the engine over on the starter for at least four cycles.
8. Write down the maximum gauge reading for cylinder number-1
9. Repeat this procedure for the seven remaining cylinders.

Notes:

- Compression readings should be uniform in all eight cylinders with less than 20 to 25 pounds variation, from best to worst.
- The actual gauge reading is not as important as finding any cylinder that is way below the average.
- The engine speed (RPM) that the starter turns the engine over factors into the compression gauge reading.
- The engine's mechanical compression ratio, camshaft duration, and other specs also affect the numbers, but should be the same in all cylinders.
- A head gasket leak can cause a loss of 50 to 70 psi.

The following chart shows some approximate compression pressure readings with the engine warm, all plugs removed, throttle open and the battery fully charged.

Engine (ci)	Gauge Pressure (psi)	Acceptable Pressure Variation (psi)
347 to 392	140	25
393 to 440+	150	25

Leak Test

To run the test, follow this procedure:

1. Remove all the spark plugs
2. Rotate the engine to TDC on the number-1 cylinder, which is the compression stroke (both valves closed)
3. Screw the test fitting into the number-1 spark plug hole
4. Connect the fitting to the leak test gauge (typically a quick-disconnect fitting)
5. Remove the radiator cap
6. Remove the air cleaner
7. Remove the crankcase filler cap or breather
8. Turn on the gauge, which applies full air pressure to the selected cylinder
9. Listen for air escaping (leaking) from carb, crankcase, headers, or exhaust manifold.; look for obverse bubbles in the radiator water
10. Repeat this procedure for the other seven cylinders

Notes:

- Be sure that the 90- to 100-psi shop air maintains its pressure during the test.
- Air escaping through the carb indicates a poorly seated or bent intake valve or a bad head gasket.
- Air escaping through the tailpipe or header indicates a leaking or burned exhaust valve or a bad head gasket.
- Air bubbling into the radiator indicates a blown head gasket or air escaping through the water pump outlets.
- Air in the crankcase (listen at the valvecover breather) indicates a bad ring or ring seal.

Basic Tests

Most engine tests are performed during or after the engine is broken-in. Unless you have switched the engine to a carburetor and distributor, the next test is to *not* set the timing because there is no distributor and the TDC and timing numbers are in the computer. That means that the crank sensor (right rear of block) must be functioning properly.

The second test is performed after the engine is running. This is the fluids and temperature test. Be sure that there are no fluids on the garage floor (gas, oil, or water) and that the engine water temperature is 180 degrees F. Hunt for vacuum leaks; it's the number-one driveability problem.

The third test is the compression test, which requires a compression gauge but they are inexpensive and almost every engine builder has one.

The compression should be tested after the engine's first running cycle or after the break-in cycle is complete.

It is helpful to keep a battery charger hooked up to the battery during the compression test to ensure that the cranking speed (RPM) for each cylinder is maintained for each cylinder.

The fourth test is the leak test. It gives you more information than the compression test, but the leak test gauge is less common and more expensive. The leak test is best run after the break-in cycle even if nothing appears to be wrong because the numbers for each cylinder can be useful if problems are encountered later. Always write down the test results in your engine build book.

An engine always leaks a small amount. One of the nice features of a leak test gauge is that it gives you a numerical answer for how severe the leak is, such as 3 percent, less than 10 percent, 15 to 20 percent, over 30 percent, etc.

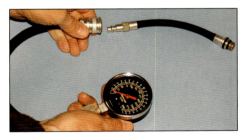

The compression gauge is the most common engine-testing gauge. It's inexpensive and easy to use. Pull all the plugs, install the connector (top right) into the plug location, and hook up the gauge.

A leak is difficult to find and requires shop air, which attaches to the opposite end of a two-dial gauge. A leak tester can give you more information about the engine than a compression test.

Similar to the compression test, all cylinders should be close to the average in leakage number. With a new engine, before being run, sometimes things stick and they fix themselves during break-in, which is one reason to do the leak test after the break-in cycle. Write down your results. Leak test gauges vary; therefore, always try to use the same gauge when you make leak test comparisons.

Air/Fuel Ratio

The air/fuel (A/F) ratio for an engine is generally said to be either rich or lean. The dyno gives you direct information on the richness level of the engine. If this level needs to be adjusted, it should be done before any fine-tuning.

The basic A/F ratio relates to the engine's BSFC, which is indicated by the dyno test. It is typically related to wide-open throttle and is typically desired to be .500, which is also used in MPI injector-size calculations.

With a carburetor-based induction system, the engine's A/F ratio is adjustable by using the carburetor's primary and secondary jets and metering rods (on Edelbrock/Carter-based designs such as AFB and AVS) or metering plates and jets (on Holley-based designs).

There are two types of fuel injection: TBI and MPI. The throttle-body version looks like a carburetor but adjusts like an MPI system, by reprogramming the computer. In some cases, the upgraded program is in the computer, especially in handheld work-with units; all you have to do is select a new map.

Tuning for Power

Once the engine starts and runs through the break-in cycle, you are ready to tune it for better operation. That first power run holds the keys to the next steps in the dyno evaluation. If the power run went smoothly and the power numbers were acceptable, you can install the engine in the car. If the run looked lean or rich (A/F ratio), an adjustment can be made.

It's a different story if the engine doesn't run properly. Most dynos have enough sensors and measuring devices that you can figure out what is wrong. It may need reprogramming to enrich or lean out the fuel mixture before you make another power run. When the engine isn't running right, try spot-checks rather than a full power run.

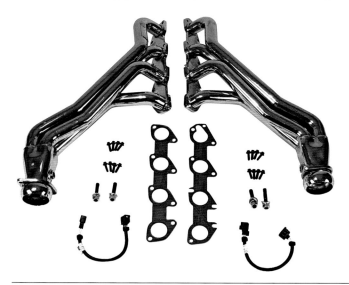

Once on the dyno, you can adjust the headers: longer tubes, bigger tubes, more or less collector. (Photo Courtesy BBK)

MPI

With MPI, there is no distributor so there is no timing to set (you must reprogram the computer to gain spark advance) and there are no jets (you must reprogram the computer to change the fuel level). Typically a fuel-spark map (as with the production baseline) has a fuel window that allows for small changes and adjusts the amount of fuel delivered to the engine based on inputs from the oxygen sensor plus the other six or seven sensors used in an MPI system.

For example, you could install a low-restriction air cleaner or cold-air system and the computer would adjust. However, if you install a CNC-ported head, it will be too lean and have to be reprogrammed.

TBI

The TBI systems that are now available are similar to MPI systems. The TBI computer typically has more than one fuel-spark map so the upgrades can be more easily accomplished by simply selecting the next upgrade map.

Carburetor

The amount of fuel delivered to the engine with a carburetor is a function of the jets, or metering plates. The typical Holley 4-barrel carb has primary jets and secondary jets. If the dyno run indicates that the engine is too lean, you install bigger jets. If it is too rich, you install smaller jets.

The Edelbrock/Carter AFB and AVS carbs also have primary jets, but they are adjusted without taking the carb apart. You first install a metering rod in each secondary jet, which can be changed without taking the carb apart. A smaller metering rod (less diameter) means more fuel into the engine. The AVS has an air door over the secondary throttle bores, which can be adjusted from outside the carb.

Nitrous Oxide

Nitrous oxide systems are common and come in many versions. In general there are plate systems, direct-port injection systems with multiple injectors, and combinations of the two. The advertised horsepower gains by the various manufacturers can be easily more than 500 hp using a race system.

However there are limitations based on the fuel. With pump gas (92 octane) or pump gas with octane booster, the horsepower gains are between 100 and 200 hp. This is considered a first-step nitrous system.

Using pump gas, the system requires less total spark advance by 2 to 4 degrees (based on non-supercharged engines). On race-only systems, this total spark advance decrease is even more dramatic.

If your engine is a naturally aspirated engine, street nitrous kits require colder spark plugs by one or two steps. For example, if you were currently using an 11-heat-range Champion-style spark plug, you would drop the plug's heat range to the 9-heat range level with a street nitrous kit. With race-only nitrous kits, the colder plug requirement increases even more.

Most off-the-shelf nitrous kits are based on the use of a 4-barrel carburetor or 4-barrel throttle body; large single-throttle body kits are available for MPI engines.

Because nitrous systems have a lot of parts (nozzles, hoses, solenoids, fittings, switches, and nitrous bottle) I strongly recommend installing a complete kit from manufacturers such as ZEX, Nitrous Supply, Edelbrock, Nitrous Express, or Nitrous Oxide Systems.

Supercharger

A belt off the nose of the crankshaft drives a supercharger. Many manufacturers produce the basic supercharger and they are available in several sizes. In general, if you plan to use a supercharger on the street you should limit the engine to 9.0:1 compression ratio or less. In addition, a basic supercharger kit should be designed for a maximum of 6 to 8 psi of blower pressure.

Supercharger Carb

For this book, I selected an Edelbrock supercharger kit because it is readily available and Edelbrock offers carburetor tips for the use of this kit on 347- to 450-ci wedge-head engines.

MPI

Ignition timing is controlled by the computer and the A/F ratio is controlled by the computer so you can't change it unless you reprogram the computer. Because of the dual-plug configuration and the reasonably small combustion chamber, Gen III Hemis do not like a lot of spark advance; 23 to 25 degrees total advance. This means that there isn't going to be much performance gain by advancing the production timing. Any performance gain comes from optimizing the A/F ratio for the specific package. This is accomplished by reprogramming the computer or the work-with device.

With a distributor and single-plug setup, the timing increasing to 35 degrees.

Supercharged-1 Conversions

Most supercharger manufacturers have recommended maps for use with their kit and you should follow them as closely as possible. Don't add things or change things without consulting the manufacturer. Remember to reprogram after each 30- to 50-hp step. Superchargers like colder plugs so you should drop down one or two heat ranges when you add a blower.

Supercharged-2 Hellcat

The supercharger package is leading-edge technology, but some improvements are still possible.

Nitrous kit manufacturers now offer large, single, round plates to install between the single, large, round throttle body and the intake manifold.

Edelbrock Supercharger Kit*	
Edelbrock's specific recommendations are as follows:	
Supercharger TBI/ Supercharger	GMC 6-71 or equivalent**
Carburetors	Two model 1405 AFBs
Primary Jets	.101 inch
Secondary Jets	.101 inch
Metering Rods	.070 x 0.042 inch
Step-Up Springs	Orange
Needle and Seat Assemblies	.110 inch (standard is .0935)
* Designed for 347- to 450-ci V-8 carbureted engines.	
** Other manufacturers that have similar-size, positive-displacement, belt-driven superchargers.	

BOLT TORQUE SPECIFICATIONS

	5.7 (ft-lbs)	5.7 Eagle (ft-lbs)	6.1 (ft-lbs)	6.4 (ft-lbs)	6.2 (ft-lbs)	426 Aluminum (ft-lbs)
Alternator (Generator) mount	40	48	40	40	48	40
Block pipe plugs						
1/4 NPT	15	15	15	15	177 in-lbs	
3/8 NPT	20	20	20	20	20	
Bellhousing to block bolts	33	33	33	50	33	33
Cam sprocket bolt	90	90	90	90	70	90
Cam tensioner plate bolts	21	106 in-lbs	106 in-lbs	21	97 in-lbs	21
Chain case cover	21	21	21	21	21	21
Lifting stud	41	41	40	40	41	40
Coil to head bolt	62 in-lbs	62 in-lbs	62 in-lbs	62 in-lbs	62 in-lbs	
Connecting rod cap bolts	15 plus 90 degrees	15 plus 90 degrees	33 plus 60 degrees	33 plus 60 degrees	30 plus 90 degrees	30 plus 90 degrees
Cylinder head bolt						
M12 (3 step)	40 plus 90 degrees	40 plus 90 degrees	40 plus 90 degrees	40 plus 90 degrees	40 plus 90 degrees	80
M8 (2 step)	25	25	25	25	25	25
Cylinder head cover bolts	71 in-lbs	71 in-lbs	70 in-lbs	70 in-lbs	97 in-lbs	70 in-lbs
Exhaust manifold screws/nuts	18	18	18	23	19	18
Flexplate-to-crank	70	70	70	70	70	70
Flywheel-to-crank	55	55	55	55	55	55
Flexplate-to-converter	31	31	31	50	31	31
Intake manifold bolt	9	9	9	9	9	9
Lifter guide holder	9	9	9	9	9	9
Main bearing cap bolt						
M12	20 plus 90 degrees	20 plus 90 degrees	21 plus 90 degrees crossbolts	21 plus 90 degrees	21 plus 90 degrees	80 crossbolts
M8	23	21	16	23	23	25
MDS Solenoid bolts				8		
Motor mount to block (front)	70	70	70	44	45	70
Motor mount, through bolt/nut	70	70	70	45	44	70
Oil dipstick tube	9	9	23	23	97 in-lbs	9
Oil filter adapter	21	21	21	21	30	21
Oil pan bolt	9	9	9	9	9	9
Oil pan drain plug	20	20	20	20	20	20
Oil pump attaching bolt	21	21	21	21	23	21
Oil pump cover bolts	95 in-lbs	95 in-lbs	95 in-lbs	95 in-lbs	133 in-lbs	95 in-lbs
Oil pump pickup tube, bolt and nut	21	21	21	21	21	21
Piston oil cooler jet bolt	10	10	10	10	8	NU
Rear seal retainer bolts	11	11	11	11	115 in-lbs	11
Rear insulator bracket bolts	50	50	50	40	50	50
Rear mount to transmission	50	50	50	24	50	50
Rear mount to bracket	50	50	50	50	23	50
Rear mount to crossmember	30	30	30	45	48	30
Rocker arm attaching bolt	16	16	16	16	16	16
Spark plug	13	13	13	13	13	13
Strut tower support bolts				28		
Thermostat housing	21	21	21	21	21	21
Throttle body screws	9	9	9	9	9	9
Variable valve timing solenoid bolt				97 in-lbs		
Vibration dampener bolt	130	130	129	127	239	130
Water pump to front cover	21	21	21	21	21	21

ENGINE CLEARANCES

	5.7 (inch)	5.7 Eagle (inch)	6.1 (inch)	6.4 (inch)	6.2 (inch)	426 (inch)
Crank end play	.002 to .011	.002 to .011	.002 to .011	.002 to .011	.002 to .011	.002 to .011
Connecting rod side clearance	.003 to .0137	.003 to .0137	.003 to .0137	.003 to .0137	.0028 to .0146	.003 to .0137
Connecting rod bearing clearance	.0015 to .0026	.0007 to .0023	.0007 to .0029	.0007 to .0029	.0006 to .0022	.0007 to .0029
Main bearing clearance: all journals	.0004 to .0016	.0009 to .0020	.0009 to .0020	.0009 to .0020	.0006 to .0020	.0009 to .0020
Piston clearance at top of skirt	.012 to .023	.012 to .023	.00096 to .002	.00096 to .020	.00167 to .0266	.00096 to .002
Piston pin clearance (in piston)	.0001 to .0005	.00023 to .00059		.0002 to .0006	.00023 to .00059	
Piston pin clearance (in rod)	Pressed	Pressed	Pressed	.0008 to .0022	.0008 to .0022	.0008 to .0022
Ring gap to compression rings						
#1	.015 to .025	.015 to .025	.0118 to .0157	.0118 to .0157	.0138 to .0197	.0118 to .0157
#2	.009 to .020	.009 to .020	.0137 to .0236	.0137 to .036	.0197 to .0299	.0137 to .0236
Ring gap to oil control (steel rails)	.0059 to .0259	.0059 to .0259	.0079 to .028	.0079 to .028	.0059 to .0177	.0079 to .028
Ring clearance in piston grooves	.001 to .0035	.001 to .0035	.0007 to .0024	.0007 to .0026	.0016 to .0028	.0007 to .0024
Oil ring	.001 to .0031	.001 to .0035	.0007 to .0091	.0007 to .0091	.0007 to .0083	.0007 to .0091
Tappet clearance in bore	.0008 to 0025	.0008 to .0025	.0007 to .0024	.0007 to 0024	.0007 to .0024	.0007 to .0024
Valve stem to guide clearance (both)				.0008 to .0025	.0008 to .0025	.0008 to .0025
Intake	.0008 to .0025	.0009 to .0024	.0008 to .0025			
Exhaust	.0009 to .0025	.0010 to .0023	.0010 to .0028			

SOURCE GUIDE

Accel/Holley Performance
1801 Russellville Rd.
Bowling Green, KY 42101
holley.com/brands/accel

AEM Performance
    Electronics
2205 W. 126th St., Unit A
Hawthorne, CA 90250
aemelectronics.com

American Racing Headers
880 Grand Blvd.
Deer Park, NY 11729
americanracingheaders.com

Arrow Racing Engines
3729 Auburn Rd.
Auburn Hills, MI 48326
arrowracingengines.com

BBK Performance
27440 Bostik Ct.
Temecula, CA 92590
bbkperformance.com

BHJ Dynamics
1651 Atlantic St.
Union City, CA 94587
bhjinc.com

Bischoff Engines
27545 State Rt. 1
Guilford, IN 47022
besracing.com

Borsh Motorsport
38000 Hills Tech Dr.
Farmington Hills, MI
    48331
bosch-motorsport.com

Brisk USA Performance
6942 FM 1960 Rd. E., #158
Humble, TX 77346
brickusa.com

Bullet Cams
8785 Old Craft Rd.
Olive Branch, MS 38654
bulletcams.com

Callies Performance
901 South Union St.
Fostoria, OH 44830
callies.com

Canton Racing Products
232 Branford Rd.
North Branford, CT 06471
CantonRacingProducts.com

CP-Carillo Industries
1902 McGaw
Irvine, CA 92614
cp-carillo.com

Charlie's Oil Pans
5281 S. Hametown Rd.
Norton, OH 44203

Clevite
23030 Mahle Dr.
Farmington Hills, MI 48335
mahle-aftermarket.com

Comp Cams
3406 Democrat Rd.
Memphis, TN 38118
compcams.com

Corsa Performance
140 Blaze Industrial Pkwy.
Berea, OH 44017
corsaperformance.com

CSR Performance Products
16936 Cty. Rd. 252
McAlpin, FL 32062
csr-performance.com

Crane Cams
1830 Holsonback Dr.
Daytona Beach, FL 32117
cranecams.com

Del West
28128 W. Livingston Ave.
Valencia, CA 91355
delwestusa.com

Diablosport
1865 SW 4th Ave., Ste. D2
Delray Beach, FL 33444
diablosport.com

Diamond Racing Products
23003 Diamond Dr.
Clinton Township, MI
48035
diamondracing.net

Dura-Bond Bearing
3200 Arrowhead Dr.
Carson City, NV
89706-2003
dura-bondbearing.com

Eagle Specialty Products
8530 Aaron Ln.
Southaven, MS 38671
eaglerod.com

Edelbrock
2700 California St.
Torrence, CA 90503
edelbrock.com

Federal-Mogul Motorsports
27300 W. 11 Mile Rd.
Southfield, MI 48034
federal-mogul.com

Fel-Pro
26555 Northwestern Hwy.
Southfield, MI 48033
felpro-only.com

Firecore Performance
Products
11430½ Station Rd.
Columbia Station, OH
44028
firecore50.com

FI Tech EFI
12370 Doherty St., Ste. A
Riverside, CA 92503
fitechefi.com

Fuel Air Spark Technology
3400 Democrat Rd.
Memphis, TN 38118
fuelairspark.com

Gibtec Pistons
333 W. 48th Ave.
Denver, CO 80216
gibtecpistons.com

Harland Sharp
19769 Progress Dr.
Strongsville, OH 44149
www.HarlandSharp.com

Hastings Piston Rings
325 N. Hanover St.
Hastings, MI 49058
www.hastingsmfg.com

Hedman Hedders
12438 Putnam St.
Whittier, CA 90602
hedman.com

Hogan's Racing Manifolds
303 N. Russell Ave.
Santa Maria, CA 93454
hogansracingmanifolds.
com

Icon Pistons
1040 Corbett St.
Carson City, NV 89706
uempistons.com

Indy Cylinder Head
8621 Southeastern Ave.
Indianapolis, IN 46239
indyheads.com

JE Pistons
10800 Valley View St.
Cypress, CA 90630
jepistons.com

Jesel Inc.
1985 Cedar Bridge Ave.
Lakewood, NJ 08701
www.jesel.com

Hi-Lift Johnson
Muskegon, MI 49442
hylift-johnson.com

K1 Technologies
7201 Industrial Park Blvd.
Mentor, OH 44060
k1technologies.com

Koffel's Place
740 River Rd.
Huron, OH 44839
b1heads.com

Kook's Custom Headers
141 Advantage Pl.
Statesville, NC 28677
kooksheaders.com

Lucas Oil Products
302 North Sheridan St.
Corona, CA 92880
LucasOil.com

Mahle Aftermarket
23030 Mahle Dr.
Farmington Hills, MI 48335
mahle-aftermarket.com

Mancini Racing
P.O. Box 239
Roseville, MI 48066
manciniracing.com

Manley
1960 Swarthmore Ave.
Lakewood, NJ 08701
manleyperformance.com

Melling Performance
P.O. Box 1188
Jackson, MI 49204
melling.com

Milodon
2250 Agate Ct.
Simi Valley, CA 93065
milodon.com

Modern Cylinder Head
586-468-7914
moderncylinderhead.com

Modern Muscle
340 C Colonel Lee Rd.
Martinsville, VA 24112
modernmuscle
performance.com

Molnar Technologies
4101 40th St., Ste. 4
Kentwood, MI 495121
molnartechnologies.com

Moroso Performance
Products
80 Carter Dr.
Guilford, CT 064437
moroso.com

MSD
1490 Henry Brennan Dr.
El Paso, TX 79936
msdperformance.com

Ole Buhl Racing
obr.uk.com

PAC Racing Springs
255 Raceway Dr.
Mooresville, NC 28117
racingsprings.com

Performance Trends
31531 W. Eight Mile Rd.
Livonia, MI 48152
performancetrends.com

Pro Gram Engineering
475 5th St. N.E.
Barbarton, OH 44203
pro-gram.com

Pro/Race Performance
Products
42295 Avenida Alvarado,
Unit 3
Temecula, CA 92590
pro-race.com

PSI
42 High Tech Blvd.
Thomasville, NC 27360
800-448-1223

Scat Industries
1400 Kingsdale Ave.
Redondo Beach, CA 90278
scatcrankshafts.com

Schumacher Creative Services
2025 N.E. 123rd
Seattle, WA 98125
engine-swaps.com

SCT Performance
4150 Church St., Ste. 1024
Sanford, FL 32771
sctflash.com

Smith Brothers Pushrods
2895 S.W. 13th St.
Redmond, OR 97756
www.pushrods.net

Stanton Racing Engines
100 Memorial Dr.
Nicholasville, KY 40356
stantonracingengines.com

Stealth Titanium Valves
2046 Depot St.
Holt, MI 48842
tricktitanium.com

Stef's Performance Products
693 Cross St.
Lakewood, NJ 08701
stefs.com

T & D Machine Products
4859 Convair Dr.
Carson City, NV 89706
www.tdmach.com

Taylor Cable
301 Highgrove Rd.
Grandview, MO 64030
taylorvertex.com

TCI Automotive
151 Industrial Dr.
Ashland, MS 38603
tciauto.com

Total Seal Piston Rings
22642 N. 15th Ave.
Phoenix, AZ 85027
totalseal.com

Trend Performance
23444 Schoenherr
Warren, MI 48089
trendperform.com

Tri-Tec Motorsports
4141 West Grand Blanc Rd.
Swartz Creek, MI 48473
tritecmotorsports.com

Tube Technologies
1555 Consumer Circle
Corona, CA 92880-1726
ttiexhaust.com

Wilson Manifolds
4700 NE 11th Ave.
Ft. Lauderdale, FL 33334
wilsonmanifolds.com

Wiseco Piston
7201 Industrial Park Blvd.
Mentor, OH 44060
wiseco.com